Estimating the
Missing People
in the
UK 1991
Population Census

Estimating the Missing People in the UK 1991 Population Census

Dr. H.M. Wasiul Islam

authorHOUSE®

AuthorHouse™ UK
1663 Liberty Drive
Bloomington, IN 47403 USA
www.authorhouse.co.uk
Phone: 0800.197.4150

Published by AuthorHouse 12/03/2015

ISBN: 978-1-5049-9420-0 (sc)
ISBN: 978-1-5049-9423-1 (e)

Print information available on the last page.

Any people depicted in stock imagery provided by Thinkstock are models,
and such images are being used for illustrative purposes only.
Certain stock imagery © Thinkstock.

This book is printed on acid-free paper.

In memory of my parents: Badrul Islam Nurnunnabi

And

Hasina Banu

Acknowledgements

I express my sincere gratitude to my M. Phil supervisor Mr. C. A. O'Muircheartaigh, Senior Lecturer, Department of Statistics, London School of Economics and Political Science (LSE), for his keen interest, co-operation, guidance over the whole duration of this study.

I also acknowledge with gratitude the support I received from Professor Ian Diamond, Chairman, Social Statistics Department, Southampton University. Thanks also go to Dr. M. Knott, Chairman, Statistics Department, LSE, for his valuable advice and to Dr. Irini Moustaki, Lecturer, Department of Statistics, LSE, for her help and support right from the beginning as a friend.

Regarding publishing the original M. Phil thesis dissertation into this book form, I have received valuable help, encouragement and practical support from Dr. Arbab Akande, a Fellow of the Chartered Institute of Profesional and Development, London; Mr Md. Soukat Ali, writer and author; Mr. Md. Shahadat Ali, Rtd. Civil Servant of Gants Hill, Ilford, London, without which this book would may not have been possible.

Finally, this book simply would never have been completed without help from home. My wife Mehbuba Shirin and son Sakkhar Islam not only provided encouragement and support in completing this book but also share all the difficulties we faced in life.

Contents

Chapter 1

Introduction

In recent years, interest in improvements in the taking of censuses of population and housing have been realized in many countries. This is due, among other things, to their growing use in formulating policies and programmes, in the dispensation of Government funds, and in planning. In 1961, England and Wales first marked the formal statistical checks on the completeness and quality of census enumeration. Some limited demographic checks, using birth and death Registration records for instance, took place immediately after the 1951 census; but these were not as extensive, nor as pre-planned, as those devised for subsequent censuses (OPCS, 1983). The best results of the 1991 Census Validation Survey suggested that the census missed about 288,000 people or 0.5 per cent of the population present in private households in England and Wales on census night. Among these people 162,000 were residents, 98,000 were visitors, while the residential status of the remaining 28,000 was not known. The CVS estimates (best) of overall undercoverage of household spaces was estimated as 198,000 -- a little under one percent of the total while the estimated overall undercoverage of households was 125,000 -- a little over 0.5 percent of the total. However, there was no statistically significant difference between the net underenumeration of different areas -- of household spaces, households, and resident households (Heady et al, 1994).

The degree of under-enumeration from the 1981 PES (Post Enumeration Survey) is of the same magnitude as the extent of under-enumeration established in the coverage checks of the 1991 CVS. From the 1981 PES it was observed that there was a big difference between London -- especially Inner London -- and the rest of the country. The same result held in 1991. Under enumeration of persons in the Inner London area is estimated to be about 2.5 per cent compared with 0.3 per cent outside London – the level in Outer London is about 1.0 per cent. Some 2.8 per cent of households were missed in Inner London compared with about 0.4 per cent elsewhere. From the U.S. census reports it is also known that i) men tend to be missed at a higher rate than women and ii) blacks tend to be missed at a higher rate than non-blacks. If the rate of net undercount were nearly constant across races, sexes, age-groups and regions, few people would be concerned. Some national and Local Government agencies rely on census figures to determine how to allocate funds and resources to various Government programs. If large demographic groups are differentially undercounted in the census, such funds and resources, which are allocated to administrative units partly or wholly on the basis of their estimated sizes, are inequitably distributed. A group with a large relative undercount receives somewhat less of these resources per capita. Moreover, all inhabitants of an area with a high proportion of members of such a group probably suffer as well.

In this situation it is very important to investigate the ability of different methods to identify the extent of undercoverage. It is needless to mention that the ability of different methods to produce accurate estimates essentially depends on accurate and reliable data. If the quality of the data collected is not good, even the best method of estimating the undercoverage will not be able to give good estimates.

The purpose of this book is to compare different methodologies for estimating census coverage and to investigate how well they work and to provide a methodology for distributing estimated missing people throughout the country. This dissertation also gives an imputation technique and develops a methodology to estimate the total error of any census and/or survey.

An Outline of the book

The book is divided into nine chapters. **Chapter 2** contains a review of the different methods of estimating the undercoverage around the world with their merits and demerits. We mainly focus on one of the post enumeration survey estimates known as dual system estimate (DSE). We describe the method in detail and also the main methodological problems of the method. We also discuss some alternative estimates of the DSE as well as the triple system method of estimation.

In **Chapter 3** we describe the 1991 U.K. Census process and methodology and sampling design of the Census Validation Survey (CVS). Some of the CVS findings were also compared with demographic estimates and we discuss the main reasons for the discrepancies between the two estimates. We also discuss the ways of handling the unresolved cases of the census by the CVS interviewers.

In **Chapter 4** we present an example with the help of hypothetical data of dual-system models to estimate the missing people as well as the undercount of the census. We present three alternative estimation procedures of the people missed by both the census enumerators and the CVS interviewers. The three estimation procedures were chosen because we believe these three estimates will produce a range of missing people under most of the data collection situations.

In **Chapter 5** we give details of the log-linear model as an extension of the dual system method to estimate the undercoverage by using information from three different sources one of which may be administrative lists. We discuss four alternative log-linear models and use hypothetical data to estimate the population total.

In **Chapter 6** we present a regression model to estimate the local authority population total. We discuss in detail why and how we estimate the dependent variable for the regression model for 403 local authorities. For evaluation purposes we use another independent estimate known as the `Goldstandard' estimate.

In **Chapter 7** we briefly describe several types of missing unit, the effect of Missing unit, different methods of imputation and our proposed methods of imputation.

In **Chapter 8** we deal with the total error model. We divide the total error into three components and describe each of these components with the complete procedure of measuring the error from each of these three components.

In **Chapter 9** we present recommendations which we believe are necessary for the improvement of the future U.K. census.

Chapter 2

Census Evaluation: Review of the Literature

2.1 Introduction

Pre-modern censuses or counts of population were carried out with some specific purpose - such as taxation or military conscription - in mind. The modern censuses are much more than headcounts: "it also implies the collection and publication of a great variety of information on the characteristics or `attributes' of the individual, such as age, sex, marital status, birthplace, economic activity, etc., and the composition of the households among which the population is distributed" (Dewdney, 1983). In 1967 the United Nations defined the census as "the total process of collection, compiling, evaluating, analysing and publishing demographic, economic and social data pertaining, at a specified time, to all persons in a country or in a well-defined part of a country".

According to Yaukey (1985), a modern census has four key elements: (1) It should be universal, that is everyone in the census area should be enumerated. (2) It should be simultaneous, that is, everyone should be counted at the same time to minimize the underenumeration or overenumeration. (3) It should be periodic, that is, everyone should be counted at regular intervals in order to permit measurement of changes in the population. (4) Finally a census should be individual, that is, the enumeration of each should include different descriptive variables about that person (age, sex, race, etc.) so that individual-level variables can be cross-classified.

Every census in every country faces a challenge to meet the above criteria. The two strongest criticisms of the present census are first that unit costs have increased significantly and second the problem of differential undercount by sex, race and region. The volume of gross error also contributes to the growing momentum and advocacy for fundamental change in the census operations.

An important indicator of census data quality is to measure the gross census error. Such measures consider not only omissions (that produce undercounts) but also double or multiple enumerations (that produce overcounts). There are several ways to measure these errors but demographic and survey estimates are the main and more popular. In the following we will describe some of the methods with their advantages and disadvantages.

2.2 Demographic Analysis

The general methods of making demographic estimates of the population for valuation purposes are based on the estimates of the components of population change, which can occur only in two ways---through reproductive change (also known as natural increase) and through migration. If the number of births (B) during a given period exceeds the number of deaths (D), the reproductive change is positive, and the population increases in size. On the other hand if the number of deaths exceeds the number of births, reproductive change is negative and the population declines. A similar situation holds for migration. If the number of immigrants (I) exceeds the number of emigrants (E), the population increases; if the reverse migration situation exists the population becomes smaller.

In countries in which vital registration system and migration data are relatively complete or accurately measured, the population estimates P_t for census evaluation based on demographic analysis are derived by the basic demographic accounting equation.

$$P_t = B - D + I - E \qquad (2.1)$$

Current population totals can also be estimated by using census counts from a previous year in conjunction with vital registration and migration data. This was the practice in the U.K. 1991 demographic estimates, in which 1981 census counts were used as baseline estimates P_b (adjusted for coverage error), with recorded births and recorded deaths data used to estimate the natural growth of the population in the last ten years and immigration and emigration data used for estimating the net effect of migration at national level.

$$P_t = P_b + B - D + I - E \qquad (2.2)$$

Population estimates from the above equation are then compared with the corresponding census counts to yield a measure of net census coverage.

Coverage error = Demographic estimate − Census count   (2.3)

In the U.S. the estimated total population P_t based on demographic analysis in 1990 involves first developing estimates for the population in various categories, such as age-sex-race groups. The particular method used to estimate the population total for the various demographic subgroups depends primarily on the nature and availability of the required demographic data. Different demographic techniques were used for the population under age 55, 55-64, and 65

5

and over; the total population is the sum of these subgroups. In the following we will discuss some of them in brief.

2.2.1 Estimation of subgroups

Age under 55

Estimates of the population under age 55 in 1990 are based directly on the estimates of the components of population change for both sexes and each race category. Births for 1935 to 1990 corrected for underregistration are carried forward to later census dates with statistics and estimates for deaths, immigration, and emigration. The population estimates P_1 are derived by the basic demographic accounting relationship:

$$P_1 = B - D + I - E \tag{2.4}$$

Age 55-64

For the population age 55-64 different analytic techniques are used to develop the demographic estimates in 1990 for this age group as there were no national data on registered births and underregistration factors for this group (i.e. birth from 1925 to 1935). Estimates for births to the white population for 1925- 1935 developed by Whelpton (1950) are carried forward to 1940 with lifetable survival rates and to 1990 with components of change to estimate the population age 55-64. Coale and Rives (1973) developed revised population estimates for the black population which were carried forward to 1990 with components of change. Estimates for the other races of the population aged 55-64 were derived from assumptions about the consistency of age patterns of coverage in earlier censuses and the use of expected sex ratios. The equation used to estimate the population of the age-group 55-64 is,

$$P_2 = T - D + I - E \tag{2.5}$$

where T is the estimate in a previous time period (1925-1935 births for White, 1960 population for Blacks, 1990 population for other Races) and D, I, and E are same as before.

Age 65 and over

In 1990 in the U.S.A. to estimate the population age 65 and over (P_3), administrative data on Medicare enrollments were used for both sexes and all race groups. It is generally presumed that the Medicare enrolment is quite complete, even then adjustments to the basic data was necessary for groups known or suspected to be omitted(i.e., persons eligible for Medicare coverage but not enrolled, aliens resident in the country for less than 5 years, certain Federal employees and annuitants). The equation used is:

$$P_3 = M + m \qquad (2.6)$$

where M is the aggregate Medicare enrollment and m is the estimate of underenrollment (Robinson et al, 1993).

Total Population

The estimated total population in 1990 based on the demographic analysis P_t represents the sum of the individual estimates for ages under 55, 55-64, and 65 and over.

$$P_t = P_1 + P_2 + P_3 \qquad (2.7)$$

2.2.2 Estimation of Components

Estimating Births

In the United States the data on births came from the vital registration system and were available at the national level only since about 1935. Correction factors were derived from the tests of birth registration completeness conducted for 1940, 1950, and 1964-68 for those years; factors for other years were developed by interpolation and extrapolation.

In the United Kingdom information on births (by sex) used in population estimates comes from the compulsory civil registration system administered by OPCS. The adjustment from birth registration to birth occurrences is made because a delay of up to 42 days is permitted between a birth and its registration.

Estimating Death

Information on deaths in the 1990 U.S. demographic estimates is based on dministrative records believed to be relatively complete. For infant deaths before 1960, it was assumed that deaths were underregistered at one-half the rate of underregistration of births; no adjustment for infant deaths was made for the years after 1960. In addition to actual deaths, life table survival rates are used to carry forward the older cohorts, to estimate the sex ratio, and to estimate migration.

Like births information in the U.K. information on deaths also come from the civil registration system.

Estimating Immigration

Immigration is the third basic factor affecting change in the population of an area and country. This factor has two major parts: legal and undocumented. Data on legal immigration may be derived from a variety of sources. In the U.S. data on legal alien immigration and adjusters are based on administrative records from the Immigration and Naturalization Service (INS). The INS data are believed to be quite complete, are timely, and require little estimation in comparison to other immigration components.

The number of undocumented immigrants is extrapolated from analyses of data on the foreign-born population obtained from the censuses and from periodic supplements to the Current Population Survey (CPS). These analyses involve a residual estimation technique in which an estimates of the legally resident foreign-born population from a census is carried forward to the survey date and compared with the foreign-born population in the survey. The difference represents the number of undocumented immigrants included in the survey (Woodrow 1992).

Estimating Emigration

The fourth and last component of the population change represents emigration of legal residents only. The volume of emigration for the 1990s based on simple extrapolations of emigration during the 1960s and 1970s in the U.S..

In the U.K. there are three types of migration estimates. These are:
a) Migration within the U.K.
b) Migration beyond the U.K. and Republic of Ireland and
c) Migration to and from the Republic of Ireland.

To estimate the net migration of the above three types of migration in 1991 information from four different sources were used. These were:

1. The National Health Service Central Register (NHSCR)
2. The International Passenger Survice (IPS)
3. Miscellaneous data on migration to and from the Republic Ireland
4. Migration data from the 1981 Census of Population

In England and Wales migration within the U.K. were estimated from the NHSCR. The movement of patients between one Family Health Service Authorities (FHSA) to another were recorded in the NHSCR. Since most people do eventually re-register with a new doctor after moving, the NHSCR figures are considered to provide a good proxy indicator of migration (OPCS, 1991). In Scotland movement of patients between one area to another are recorded by Area Health Board (AHB), while in Northern Ireland record of patients movement are kept by the Common Service Agency (CSA).

To estimate the net migration beyond the U.K. and Republic of Ireland in 1991 information on tourism and the contribution of travel expenditure to the balance of payment from The International Passenger Survey (IPS) were used. The IPS is a continuous sample survey of passengers conducted by OPCS and covers the principal air and sea routes between the U.K. and overseas. The proportion of passengers sampled varies between 0.1 to 4.0 per cent according to route and time of year (OPCS, 1991).

To estimate the migration to and from the Republic of Ireland, data from different sources were used, such as, IPS, and NHSCR; results of the Labour Force Survey (LFS) and information provided by the Government of the Republic of Ireland by the Migration Analysis Unit (MAU) of OPCS.

2.2.3 Merits and Demerits

The main advantage of the demographic estimates is that the data used to estimate various components are drawn from sources essentially independent of the census being evaluated. The data are corrected for various types of errors and, as such, are assumed to be more accurate than the census being evaluated. Perhaps, the major strengths of the demographic method are the internal consistency and interrelationships of the underlying demographic variables and the data used to measure them. A further advantage is that the particular method, data and assumptions used in demographic analysis are not fixed. As the new methods develop over time and new data or information become available, they can be incorporated into the estimates, with the hope of improving the estimates. Similarly, the assumptions of the demographic estimates may also change with the new information and hence may change the demographic estimates of the population and the coverage of a particular census over time (Robinson et al., 1993).

The main methodological weakness of demographic analysis is that the data may be incomplete or inaccurate, particularly, the immigration and emigration data (Edmonston and Schultze, 1995). The second shortcoming is that demographic analysis only provides an estimate of net census errors; it does not identify the separate effects of omissions, duplications or erroneous inclusions, and reporting errors (age, sex, race) in the census (Fay et al., 1988). Finally, demographic estimates of coverage are typically produced only for the national level (Ericksen and Kadane, 1985).

Demographic analysis methods have been used as a coverage error methodology for a long time in many countries around the world. The U.S. Bureau of Census has been using demographic analysis as a tool for coverage evaluation to assess the completeness of coverage in every census since 1960 (U.S. Bureau of Census, 1988). In 1990 demographic analysis was used not only to evaluate the completeness of the coverage of population in the 1990 census but also to evaluate the overall quality of the national estimates of net coverage based on the 1990 post Enumeration Survey (Robinson et al., 1993).

In the 1991 census of the U.K., demographic estimates were used as a check on the aggregate accuracy of Census Validation Survey estimates of undercount by age and sex group (Heady et al, 1994). They accepted the demographic estimates as more accurate estimates at the national level on the basis that most of the components of demographic estimates are quite accurately determined. In addition, demographic estimates of sex ratios have been used to adjust the regional population.

2.3 Post Enumeration Survey (PES)

The major alternative to demographic analysis is to use surveys to measure census coverage. In this method an independent sample is drawn and matched to the original enumeration on a case by case basis. A second sample of enumerated persons is drawn from the census to determine whether they are counted correctly. These matching results are then used to give an estimate of the population size either by the dual system method of estimation or by others. A comparison of the census with the estimates of population size yields the net undercount rate.

The main advantage of the post enumeration survey estimates is that it provides estimates for levels of geography below the national level and for race/ethnic groups. One disadvantage of the Post Enumeration Program is that estimates may be subject to correlation bias as people missed by the census may also tend to be missed by the PES. Another disadvantage is that the matching between two independent lists, the PES and the census, requires a substantial amount of time and money.

2.3.1 Dual-System Estimate (DSE)

In 1949 Chandra-Sekar and Deming developed a method known as the dual-system estimation method (DSE) (also known as C-D technique) to estimate undercount /overcount of population in a census. In this method data from an independent sample survey were used to estimate population coverage in the census. The model was originally developed for use in biometric studies by Peterson (1896) to estimate the size of a closed population, generally called capture-mark-recapture (CMR) technique. Chandraborty (1963) and Das Gupta (1964) extend this approach to situations involving three or more sources of information.

The approach is also used widely in other types of population. Notable among them are Geiger and Werner (1924) -physics; Lincoln (1930) -wildlife; Jackson (1933) --tsetse flies; Schnabel (1938) --fish in a lake; Dowdeswell, Fisher, and Ford (1940) --butterflies on an island; Wittes and Sidel (1968), Wittes, Colton and Sidel (1974) --epidemiology; Sanathanan (1972) -particle scanning in hysics; Blumenthal and Marcus (1975)-life testing; Green and Stollmack (1981), Rossmo and Routledge (1990) -crimes and criminals.

In the census evaluation application, the theory of the DSE is exactly the same as that of Peterson (1896), except that a human population requires slightly different assumptions. For a C-D estimator, the first capture is the census count and the second capture is the sample count. Sample data are matched with census data, on a case by case basis. The outcome can be distributed as in Table 2.1 below, displaying the fact that some people are counted by both systems x_{11}, some by one or the other but not both x_{12} and x_{21}, and some by neither x_{22}.

Table 2.1. C-D Model for Population Estimate

Census	Sample (PES)		
	In	Out	Total
In	$x_{11} = M$	x_{12}	x_{1+}
Out	x_{21}	$x_{22} = z$	$x_{21} + x_{22}$
	$x_{+1} = \check{N}_p$		$x_{++} = \check{N}_T$

where

x_{11} is the number of people in the sample matched to the census

$x_{1+} = (N_c - G - E - D - I)$

N_c is the census count

G is the number of persons incorrectly located geographically in the census

E is the number of persons incorrectly enumerated in the census (fabricated or not in sample)

D is the duplicate enumeration in the census

I is the number of persons who are enumerated in the census but have insufficient information for matching

x_{12} is the number of persons counted in the census alone

x_{21} is the weighted number of persons counted in the sample alone

x_{22} is the number of persons missed by the both procedures

$x_{+1} = \check{N}_p$ is the weighted total of the sample

$Z = x_{22} = K(x_{12}x_{21}/x_{11})$, K=1, when two methods of data collection, census and sample survey are independent

$\check{N}_T = x_{++} = (x_{1+}x_{+1}/x_{11})$ is the estimated total population

11

Assumptions of the C-D Estimator

The main assumptions of the capture-recapture approach on which the accuracy of the estimates depends are as follows:

1. **The closure assumption:** Population U is closed and of fixed size N.

2. **Autonomous independence:** The two methods of data collection - census and survey - are independent of each other. That is, the number of distinct persons enumerated in the census is independent of the number of distinct persons numerated in the survey. Each trial corresponds to a member of the true population U.

3. **The matching assumption:** It is assumed that clerical matching has occurred without error. That is, it is possible to make a determination, without error, of which individuals recorded in the sample survey are present in the census and which are not.

4. **Spurious event assumption:** It is assumed that both the census and sample survey data are free from erroneous enumeration. This means that all errors are avoided in recording both the census and the survey results.

5. **Nonresponse assumption.** A small non-response is thought to remain in the census. Sufficient identifying information is gathered about the non-respondents in both the census and the sample survey to permit exact matching from the survey to the census.

Bias may arise in the C-D estimates if the above assumptions fail, that is, if the population under study is not closed (that is, if births, deaths, or migration occurred between the first and second sample operations); if the probability of an event being recorded by one source is influenced or altered by the probability of inclusion in the other source; if the probability of inclusion in one source varies from individual to individual; if the matching of the individuals in the second and first sources is not perfect. To avoid correlation bias Chandra-Sekar and Deming suggested dividing the population into small homogeneous groups and then estimating the population parameters by combining all the sub-group estimates.

Just after the publication of the C-D technique Shapiro (1950, 1954), used the technique to evaluate the 1940 completeness of birth registration estimates in the United States. Results indicate little gain due to stratification (i.e., estimated total number of missing persons does not increased significantly). However, the most notable achievement of the evaluation test was that this was the first in which an attempt was made to formulate and write out explicit matching rules to recognize that both erroneous matches and erroneous nonmatches can occur. Davis et al (1991) points out that the bias in the overall C-D estimate attributed to matching error is determined solely by the size of the net matching error. Srinivasan and Muthiah (1968) tabulate separately the number of erroneous matches and erroneous non-matches obtained from the alternative matching criteria to estimate matching error.

Coale (1961) proposed matching independent sample survey information with recorded sources of the same events, especially in developing countries where vital registration is frequently non-existent, unreliable, or far from complete for obtaining accurate vital statistics. Since then, many countries all over the world specially in Asia, Africa and America have used the C-D technique to study population growth rates and/or to evaluate the coverage of civil systems (El-Khorazaty, 1977).

Lauriat (1967) gives a comparative study of the application of the C-D technique in Pakistan, Thailand and Turkey. The author concludes that the implementation of the technique, especially in developing countries, poses additional burdens. Seltzer (1969) discusses applications in Asia and the gains obtained due to stratification as suggested by Chandra-Sekar and Deming. An extensive description of the technique and its limitations is given by Linder (1970) while Abernathy and Lunde (1970) discuss the early history and gives detailed presentations of applications in different countries. Seltzer (1969) gives a description of a carefully executed and efficiently administered national C-D study to obtain basic demographic estimates for Liberia as a whole and separately for the urban and rural sectors of the country. The administrative setting for Pakistan, Thailand, Turkey and Liberia is described in Wells (1971). The Agency for International Development stimulated the establishment of an International Program of Laboratories for Population Statistics (POPLAB) to "establish long-range cooperative programs of work with institutions in various countries" (Linder, 1971).

Questions about the usefulness of the C-D technique in estimating events missed by both sources are raised by Brass (1971, 1975). He argues that because of the independence assumptions, the C-D technique detects only a small part of the omissions. He also states that biases and sample errors are too large for useful measures to be estimated at an acceptable cost. Thus, he suggests using easier and more economical single system designs that rely on repeated surveys. Greenfield (1975, 1976), and El-Sayed Nour (1982) on the other hand, suggest a revised procedure for the DSE to achieve more accurate results. Greenfield's procedure implies the adoption of an efficient subsample scheme rather than complete duplication of records while Nour's procedure takes into account the lack of independence of the two methods of data collection.

The U.S. Bureau of the Census introduced the use of a sample matched to the census records for coverage evaluation from the census of 1950. This approach is now known as the Post Enumeration Survey (PES) approach. After the 1980 census, considerable attention was given to correcting the census for differential undercount by sex, race and region by using extensions of the C-D technique. Ericksen and Kadane (1985) suggested using the C-D technique with a known population total and using regression models to estimate the undercount/overcount of subareas or/and subgroups while Freedman and Navidi (1986) criticise the models by saying that neither do they make explicit the assumptions of the models nor do they give any empirical foundation for them. Wolter (1986) gives alternative models for estimating coverage error in surveys and censuses of human populations with details of the assumptions of the models. Wolter (1991), Freedman (1991), and Fienberg (1991) also described their views about the problem of using the C-D technique in a Post Enumeration Survey to estimate the undercount/overcount rate of the census and suggested different corrections of the methodology. Detailed methodology including the sampling plan, treatment of nonresponse and erroneous enumeration for the application of the C-D technique in the Post Enumeration Survey of 1990 was given by Hogan (1992).

Mulry and Spencer (1991) present a methodology for estimating the accuracy of dual system estimates of population. The error or total error in a statistic is the difference between the statistic and the true value which is unknown. However, one can use an external `goldstandard' if available to estimate the undercount from the difference of the DSE and standard. In most cases no such standard exists but demographic estimates could be used to fill the gap even with their limitations. Mulry and Spencer's approach to estimate the total error in the DSE is to try to identify all the sources of error, estimate their magnitude and study their propagation through the estimation process. They divided the total error into three components such as sampling error, model error and measurement error with the assumption that all the components of error will fall into one of these three components. They sub-divided each of these three components into several other components and try to estimate the magnitude of each of these error components separately. Finally they add up all the components of error to obtain the total error.

Wolter (1986) and Childers et al (1987) classified the independence assumption of the DSE into three components:

1. Causality: the probability of an individual being included in the sample is not altered by inclusion in the census. Causal independence fails when an individual's capture history in the census alters the probabilities of capture in the sample. The estimator $\check{N}_T$ is biased downward when the odds of capture in the sample are increased as a result of capture in the census, and is upward biased when the odds of capture in the sample are reduced as a result of capture in the census (Childers et at, 1987).

2. Homogeneity: within an estimation cell, the probability of inclusion in the census is equal for each individual, and the probability of inclusion in the sample is equal for each individual. When this assumption fails, the resulting bias (called the heterogeneity bias) is generally thought to be a downward bias.

This is because individuals with a high probability of capture in the census also tend to have a high probability of capture in the sample, and conversely, individuals with a low probability of capture in the census also tend to have a low probability of capture in sample.

3. Autonomy: each individual acts alone as to inclusion in the P sample and the census. This assumption fails when, for example, household members act together in creating the census or sample enumerations (Childers et al, 1987). Cowan and Malec (1986) observed that when the other two independence assumptions hold failure of the autonomy assumption has little effect.

The most useful descriptions of the DSE, its background, assumptions, design, problems and matching procedures are given by Marks, Seltzer and Krotki (1974), and U.S. Bureau of Census (1985), while Carver (1976) and Fienberg (1992) published bibliographies on dual-system methodology.

Methodological Problems of the C-D Technique}

In the census evaluation programme the dual system estimation procedures depend on data obtained through two different information sources -usually the first source is the census enumeration which covers the whole population while the second covers a sample of it. It is quite impossible to count the population in the census and in any population survey without errors. This means that all sources of error should be thoroughly examined. In the dual record system we are mainly concerned about the following errors and discuss only them.

1. Coverage error
2. Matching error
3. Modelling error
4. Sampling error
5. Non-Sampling error

Coverage Error

Coverage error can be divided into three categories:
i) Omissions
ii) Duplications and
iii) Erroneous inclusions

Omissions

By omission we mean the non-coverage of events that belong both to the area and to the time period specified (Marks, Seltzer and Krotki, 1974). Omissions occur in two ways. In the first way census enumerators miss the entire housing unit, household, or person having no established place of residence and in the second one or more persons within enumerated housing units or households are missed. In the case where an entire housing unit is missed, it follows that all households and persons residing within the housing unit will also be missed by the census.

The possible causes of omission of housing units are imprecise boundaries of geographic or census administrative units, faulty maps, or simply coverage errors made by field staff in the pre-census listing operation, or as a result of an imprecise definition of census assignments. The households can be missed in any of the following cases:

1) when all of the members of the households were at another place at the time of the census enumeration.
2) temporarily absent during the hours of census enumeration and
3) in transit either within or outside of the country during the enumeration period.

Duplications

Duplication occurs when housing units, households, or persons are enumerated more than once. The main reasons for duplication are:

1) overlapping of enumerator assignments, due to errors made during pre-census listing
2) inability of enumerators to identify the proper boundaries and
3) some people may have more than one residence and be counted in more than one place.

Erroneous Enumerations

Erroneous enumerations are defined as persons, housing units, or households that were enumerated when they should not have been, or enumerated incorrectly in the wrong place. Examples of erroneous inclusions are persons who died before (or were born following) census day or were counted in the wrong geographical location, and persons who never existed but were counted by an interviewer. Erroneous enumeration in the census can occur at any stage of the census, but research shows that it is more frequent in the later part of the census counting i.e. during the time of coding, tabulating etc. (Wolter, 1991).

Net coverage error occurs by the simultaneous occurrence of omission, duplication and erroneous inclusion. The C-D technique attempts to minimize the extent of omission or non-coverage. However, the C-D technique may underestimate the total number of events if x_{1+} or x_{+1} or both include too few events (Ericksen and Kadane, 1985 and Wolter, 1986). Similarly, the C-D technique can overestimate because of the inclusion of too many duplicate or erroneous events in either x_{1+} or x_{+1} or both.

Chandra-Sekar and Deming (1949), Coale (1961), Seltzer and Adlakha (1969), Hogan (1992) and many others suggest a field re-investigation for non-matched events to detect and eliminate events wrongly recorded. The U.S. Census Bureau in both 1980 and 1990 (Hogan, 1992) and the U.K. Census of Population in both 1981 and 1991 (Heady et al, 1994) used a sample from the returned census forms to detect erroneous enumeration in the census evaluation surveys.

Matching Error:

The model assumes that matching persons and units is done perfectly between the two sources. That is, two types of error are assumed not to occur: erroneous matches and erroneous nonmatches. When the difference between these two types of error is zero, matching errors have no effect on the bias of the C-D estimate (Mark et al, 1974). By 'erroneous match' we mean false matches of nonmatching cases; an erroneous nonmatch means false nonmatches of matching cases.

In practice, the matching algorithm uses name, address, age, sex, race, and ethnicity. Some of the data are inaccurate, on the sample side as well as the census side. There is variation in spelling, and some persons give fictitious names. Demographic characteristics (even sex) sometimes appear to change from one interviewer to another (Freedman, 1991). The U.S. census evaluation of 1990 does not require an exact address match. Jaro (1989) developed a computer matching system which was used for the first stage matching on the individual characteristics and address information. Details of the matching process are discussed by Hogan (1992).

Several types of error may affect the C-D estimate due to imperfect matching. These are:

1. There may be duplicate or multiple enumerations in the census
2. In the census, the housing units listed may be enumerated correctly but allocated to the wrong geographic area
3. Members of a housing unit may be enumerated at the wrong location or may not be in scope for the census (i.e., should not have been counted at all)
4. Members of the housing units may be incompletely enumerated so that there is insufficient identifying information for an individual (U.S. Census Bureau, 1985).

To minimize these errors a second sample generally known as the E sample, was drawn directly from the census for the same area and using the same stratification to measure or counter the effect of each of these factors.

Model Error or Error due to the Failure of Independence assumptions

Another important assumption of the dual system estimator is that the two methods of data collection are independent. That is the model assumes that the chance of an event being recorded by the census does not influence the chance of being recorded by a sample survey. In reality the assumption of independent collection is unacceptable. Mulry and Spencer (1991), Childers et al (1987), Wolter (1986), Cowan and Malec (1986), El-Sayed Nour (1982), Greenfield (1975), Jabine and Bershad (1968) and many others argue that, in particular where the source of data is a human population, there are many possible reasons for which data can be missed systematically by both methods of enquiry. This association between the results of the two methods may be measured by an index r,

$$r = A(x_{11}x_{22} - x_{12}x_{21})$$

where A is an appropriate constant. In the Chandra-Sekar and Deming estimator this r is assumed to be zero. However, based on real observations, Chandra-Sekar and Deming (1949), Jabine and Bershad (1968) and Greenfield (1975), became convinced that the association between these two collection systems is positive.

To deal with the association bias, Chandra-Sekar and Deming, suggest that their method will give better results if it is applied to homogeneous sub-groups of the data and that the total estimate be obtained by building up from these sub-groups. The underlying argument is that if the association for each sub-group is near zero, while the association for all sub-groups combined is not zero, then a less biased estimate of x_{22} will result. El-Sayed Nour (1982) did not agree with this suggestion and argued that a better way of dealing with the association bias in estimating x_{22} is to make assumptions concerning the value of the association index r. He illustrated his logic with the following example. Consider Table 2.1 and let $x_{12} = x_{21} = x$, for simplicity. Assume the table is partitioned into two sub-tables,

$$\begin{bmatrix} M_1 & V_1 \\ V_1 & X_1 \end{bmatrix} \quad \text{and} \quad \begin{bmatrix} M_2 & V_2 \\ V_2 & X_2 \end{bmatrix} \tag{2.9}$$

where

$$(M_1 + M_2) = x_{11}, \; (V_1 + V_2) = x \text{ and } (X_1 + X_2) = x_{22} \tag{2.10}$$

Then

$$x_{22} = (x^2/x_{11}) = (V_1^2/M_1 + V_2^2/M_2) - (M_1V_1 - M_2V_2)^2/\{M_1M_2(M_1 + M_2)\} \tag{2.11}$$

This indicates that any partitioning leads to an estimate of x_{22} at least as large as that from the total table. Equation(2.11) is a result of the mathematical form of the C-D estimator, and occurs regardless of the form of the association in the contingency tables. In fact, the magnitude of the increase in the values of x_{22} due to partitioning seems to depend on the relationship between the sub-tables rather than on the relationship within the sub-tables.

Greenfield (1975), Ericksen and Kadane (1985), Wolter (1986), and Childers (1987) argue that although the method of sub-grouping offers an improved estimate, it still suffers from the defect that independence within sub-groups is assumed. Greenfield, therefore, proposed that the C-D estimate of the number of missed events should be regarded as a minimum estimate. Greenfield (1976) also suggested an alternative estimate for the upper limit, that is, an estimator for maximum missed events.

Sampling Error

The error which arises because a sample is being used to estimate the population parameters is termed the sampling error. Whatever may be the degree of care in selecting a sample, there will always be some error in the estimate. In a census coverage survey, sampling error arises because information is not collected from the entire target population, but rather only from some portion of the population.

Nonsampling Error

Besides sampling error, the sample estimate may be subject to other errors which, grouped together, are termed non-sampling errors. The non-sampling errors can occur at any one or more of the stages of a survey, i.e., planning, field work, and tabulation of survey data. Here we will discuss in short the implications of such errors. Cochran (1977), and many others, for simplicity, classified these errors, in three categories: i) Non-Response ii) Response and iii) Tabulation Error.

Non-Response Errors:

The term non-response is used to refer to the failure to measure some of the units in the selected sample. This happens mainly due to the use of faulty frames for the sampling units, biased methods of selection of units, inadequate schedules, etc. When the sampling frame is not updated or when the old frame is used on account of economy or time-saving, serious bias may occur as the targeted population is not enumerated. For example, in the PES if the old list of households is used for selection of the sample, some newly added households will be out of the sampling frame. Similarly, a number of households already demolished will remain in the frame. Thus the use of such frames may lead either to inclusion of some units not belonging to the population or to omission of units belonging to the population. In some situations, a part of the sampled units may refuse to respond to the questions or may be not-at-home at the time of interview. These may all lead to error. The main sources of these errors may be assigned as the following:

1) omission or duplication of units
2) not found at home, even after repeated calls
3) refused to give information
4) merely fails to take the trouble to return the Questionnaire and
5) is unable to furnish the information.

Response Errors:

Response errors may arise from respondents who do not make an accurate answer or may give biased answers or from the questionnaire, from the execution of the field work or from the nature of the data collection process (O'Muircheartaigh, 1977). The measurement device or techniques may be defective and may cause observational error. The main sources of these errors may be assigned as:

1) plain honest accidental mistakes in responding
2) illegible entries
3) failure of memory
4) memory-bias
5) guessing, made necessary through lack of records
6) unwillingness to give the right answers
7) refusal to give the right answer and
8) wrong answer arising from pride, called prestige-bias

2.3.3 Alternative Estimation

In dual system estimation it is assumed that the probabilities of enumeration are the same for all members of the population. From experience it is known that the probability of being enumerated in the census varies by age, sex, race and geographic area (Hogan, 1993). Following Chandra-Sekar and Deming (1949), the U.S. Bureau of the Census attempts to use post- stratification in the PES to define subsets of the population that would have constant enumeration probabilities both in the census and the survey. Nevertheless, some amount of residual heterogeneity probably remains (Alho et al, 1993).

Alho (1990) and Huggins (1989, 1991) generalized the dual system estimate to a situation in which every individual can potentially have a different probability of enumeration, which is assumed to depend on a set of independent variables through a logistic regression model. Conditioning on the observed individuals, maximum likelihood estimators (MLEs) of the parameters of the regression models, and hence of the probabilities of enumeration, can be calculated.

Alho, Mulry, Wurdeman, and Kim (1993), used conditional logistic regression to estimate probabilities of enumeration in the census and PES. This method potentially permits every individual to have different probabilities of enumeration, which are assumed to depend on a set of independent variables through a logistic regression model. The technique generalizes the stratification approach of Chandra-Sekar and Deming the same way as ordinary regression analysis generalizes fixed-effects analysis of variance, and as such it promises at least some improvements in similar future surveys. To assess the possible presence of the residual correlation bias of a post-stratified estimator, the logistic regression approach can be used by including such covariates into regression that have not been used in the definition of the PES post-strata.

The logistic approach has two limitations in the estimation of heterogeneity when the sample is drawn from a closed population with no data error. First, the logistic approach is capable of modelling observable heterogeneity only. This means that only such heterogeneity that can be "explained" by variables observed in the census can be modeled. If the true causes of heterogeneity are unobservable, then any estimates we get may be biased (Alho, 1990). Second, Alho (1990) showed that if large values of the regressors are sufficiently rare, then the conditional maximum likelihood estimates are consistent, and $\check{N}$ is a consistent estimator of N ($\check{N}/N \to 1$ as $N \to \infty$). Otherwise, the estimator may not be consistent. Alho also showed that the consistency of $\check{N}$ may fail, even if the parameter vectors are known.

Another way of dealing with the correlation bias in the dual system estimator is to replace the independence assumption by an alternative assumption which may or may not require another data set. Greenfield (1975) proposed that the C-D estimate of the number of missed events should be regarded as a minimum estimate and gives a second estimate (Appendix 1) which he proposed can be regarded as a maximum estimate of the value of omissions. He failed to support his estimate theoretically but argued that there is some empirical evidence to support the assumption proposed as providing an upper limit together with some intuitive arguments in its favour.

He developed his estimate from the correlation index r_x which can range from +1 to -1. He argued that in the field of demography, extreme values of r_x are most unlikely and even, for certain regions of the upper range, virtually impossible. He therefore, assumed a symmetrical and unimodal distribution of r_x and measures the expected value of r_x which is located at the mid-point of its possible range. This expected value of r_x is the estimate of the maximum events missed.

Greenfield (1976) also suggested another ratio estimate of the missing events. He proposed to undertake a random sub-sample from the main recording system rather than complete duplication of records to match with the main sample. He derived the estimate from the approximate variance of the ratio estimate given by Yates (1960), on the assumption that the sub-sample and main sample are simple random samples. The advantage of the proposed system is that it should permit more careful checking of the matched events, field verification of discrepancies, and probably will save costs together with the other usual advantages of smaller scales surveys over large scale surveys.

El-Sayed Nour (1982) presented an alternative approach to the estimation of missing events (appendix 2). This approach preserves the basic features of the data collection process including the lack of independence between the results of the two collection procedures. He developed his estimate by assuming that the two data sources are positively correlated and that the probability that a single event selected randomly from the population will be recorded by a given collection procedure is larger than 0.5.

Wolter (1990) developed an estimator based on an assumption of known sex ratio and an assumption of independence for females only. Bell (1993) extended the methodology and applied it to estimating correlation bias in the 1990 PES using the results from 1990 demographic analysis. An alternative method (O'Connell, 1991, O'Connell, Bloomfield, and Pollock, 1992) describes the lack of independence in terms of internal and external constraints.

Zaslavsky (1993) has developed composite estimators which use the census, the PES, and the PES Evaluation data to produce accurate estimates of population size as measured by weighted squared-or absolute-error loss functions applied to estimated population shares of domain. Several procedures were reviewed that chose between the census and the DSE using bias evaluation data or that average the two with weights that were constant across domains. The domains may be defined by geography, or by demographic factors (such as race), or by a combination of the two.

2.3.4 PES in the U.K.

In the U.K. PES (1991) which is known as Census Validation Survey (CVS), six different samples were drawn from each of the selected EDs (Enumeration Districts) or ED workloads. These EDs were selected from the selected CDs (Census Districts) with probability proportional to the estimated population size according to the 1981 Post Enumeration Survey. Out of these six different samples five samples were drawn to find missing persons and one from the counted persons to check whether the enumeration was done correctly or not. To derive the estimate of undercount they first estimated the total persons found from each of the six samples by multiplying by the inverse of their chance of inclusion in the samples and then added these six figures to build up the total figure. (More details are discussed in chapter 3).

2.3.5 PES in the U.S.

The 1990 U.S. Post-Enumeration Survey (PES) was designed to produce a census tabulation of states and local areas corrected for the undercount or overcount of the population. The PES measured Census omissions by independently interviewing a stratified sample of the population. This independent sample is known as the P sample which was a block cluster composed of either a block or a collection of blocks. Census erroneous enumerations were checked by a dependent reinterview of a sample of Census records and by searching the records for duplicates. This sample is known as the E sample and was sampled from the same block cluster as that of P sample. This sample was also used in the field to determine the extent of fictitious enumerations, inclusions by the Census of people born after the Census reference day, and the extent to which people were counted in the wrong location. A dual-system estimator was used to prepare estimates of the population by post-strata. The post-stratification of the population was based on geography, race, origin, housing tenure, age and sex. There were in total 1392 post-strata. Adjustment factors were computed as the ratio of these estimates to the census count. These factors were smoothed using a generalized linear model, and then applied to the Census counts by block and post-strata to produce adjusted census estimates.

2.3.6 PES in Bangladesh

In Bangladesh, immediately after the 1981 census, a post enumeration survey was conducted to estimate the coverage and content errors of the census at the national level and separately for urban and rural areas. The sample design was a stratified systematic sample. The population was first stratified by urban and rural areas. At the second stage, enumeration areas (EA) were arranged according to their geographic codes. A systematic random sample of EAs was then selected from this order for the coverage check.

All the sampled EAs were re-enumerated and the operation was conducted in two stages: a PES-A field survey and PES-B field follow up operation that was also used to estimate erroneous enumeration. Each EA was independently matched twice by two different persons and the results were verified by a supervisor through the use of field revisits. Non-matches were also verified in the field.

2.3.7 Triple system

The triple system is an expansion of the dual system estimate of census coverage using Post Enumeration Survey (PES). In this model an additional source of information such as administrative records is used to estimate the missing events. Zaslavsky and Wolfgang (1990) presented various triple-system estimates of the number of uncounted people, including the unrestricted no second-order-interaction model, based on a log-linear model or log-linear like models (Darroch et al, 1993) applied to a full three system table or to various marginal subtables of a three system table. They argued that "the validity of triple system estimates depends not only on how well the adopted model fits the unknown reality but perhaps more dramatically on how accurately persons are assigned in three ways: (1) in or out of scope, (2) into post-strata, and (3) into the cells of the cross-classification of sources".

Darroch, Fienberg, Glonek, and Junker (1993) developed another triple system estimator which allows for various probabilities of capture through individual parameters using a variant of the Rasch model from psychological measurement situations.

The advantage of the triple system is that the alternative independence assumption reduces the potential problem of correlation bias and the disadvantage is that the matching among three independent lists adds complexity and increases the time and resources needed.

2.4 Administrative Record Match (ARM)

In the administrative record match evaluation process a megalist of addresses was created from the several available administrative lists without duplication (unduplication). A sample from this list is drawn and matched to the census population to identify persons missed by the census. An example of a record match (possible mainly in the U.S.A.) is the determination of whether an adult whose name is obtained from a Motor Vehicle driver's license records was enumerated in the census. The percentage in the sample not matched is a measure of the census coverage error.

The main advantage of using administrative lists is that they do not rely on surveys or previous censuses. Therefore, there is not the problem of correlation bias as well as sampling errors. Also, output from an administrative records census could be produced in a timely fashion and in a manner similar to that from a conventional census. Detailed information could be produced for geographic areas, including small geographic areas with geographically referenced data.

The limitations of administrative records include the accuracy of address information and the paucity of demographic and socioeconomic variables that are included. Also there is no guarantee that the lists would cover the entire population or subpopulation of interest. Furthermore, if there is reliance on several systems of administrative records, then the records must be unduplicated by matching techniques which may be difficult because people use different addresses for different purposes. Finally, to use the administrative record, it should be updated on a continuous basis.

Some European countries, mostly in Scandinavia, such as Sweden and Finland, have conducted censuses that combine administrative records use with traditional enumeration in order to improve the quality of the census counts (Edmonston et al, 1995). Outside Scandinavia, only the Netherlands has conducted a census based primarily on administrative records.

In the U.S.A., on several occasions administrative records were matched with the census counts for evaluation. The 1940 census was matched with draft registration records for the first time by the Bureau of Census. The matching result shows that there were more males registered for the draft than enumerated in the census. The evaluation studies of the 1960 census were designed to measure omissions of persons in the census by matching the administrative records with the census counts (U.S. Bureau of Census, 1960). Record match results were based on sample studies of four population groups (1950 census count, 1950 PES, whose combined representation was believed to cover 98 percent of the population). From each of the four population groups, a sample of persons was selected, on a probability basis, and an effort was made to determine whether each person was enumerated in the 1960 census. To evaluate the coverage of the 1980 census, the post enumeration programme included the April and August 1980 Current Population Survey (CPS) samples and a sample of census enumeration. These three samples were then matched with the census counts to estimate the coverage and net coverage error of the 1980 census by dual-system methods.

2.5 Reverse Record Check (RRC)

In the reverse record check evaluation programme of the census a sample (random) of population is drawn from records created prior to the census. This record contains all persons who should be enumerated in the census and is generally built up from the previous census records, persons missed in the previous census, birth and immigration registration (Gosselin et al, 1978). These sample addresses are then traced during the census period and matched to the census to see whether or not the selected person was enumerated. The proportion of the sample which is unmatched provides an estimate of the proportion of the population which was missed in the census.

The main advantage of the method is that person changes with time can be found easily. For example, some people are very mobile during their late teens and early twenties but are less mobile as children and old adults (Mulry and Dajani, 1989). Moreover, as it is possible to draw a sample several years before the census, it may be easier to include hard-to-enumerate groups in the sample. The disadvantages of the method are that it is costly and very difficult to create a list without any error. The persons missed in the previous census would not be included in the created records and there are always some problems to trace all the sampled persons. Moreover, in order to measure the erroneous enumeration the RRC has to supplemented with a separate sample from the census.

Statistics Canada have been using the Reverse Record Check method since the 1961 Census. They started the method on an experimental basis and drew small samples from the 1956 census records. In 1966 they extended the methods and included a sample from birth and immigration registration along with a sample of 1961 census records. Statistics Canada still use this method with more accuracy and efficiency not only at the national level but also at the province level.

The U.S. Bureau of Censuses used the Reverse Record Check method to estimate the number of persons missed by the 1960 census (U.S. Bureau of the Census, 1964). The CPS-Census Retrospective Study traced CPS sample from 1976, '77, '78, '79, and '80 to the time of the 1980 decennial census and matched them to the census (Diffendal, 1986).

2.6 Multiplicity

In this method a network sampling technique is applied during census enumeration. Respondents are asked the name and addresses of their relatives, such as parents, siblings, and children. These addresses are then checked to find out whether the persons are enumerated or not. The undercount rate is estimated by the number of persons added.

An advantage of the method is that it may identify people who are hard-to-enumerate because they had loose ties to a household. A disadvantage is that the respondent often does not know the addresses of their relatives, even if they know where they live.

2.7 CensusPlus

The U.S. Census Bureau planning to introduce a new technique of evaluating census enumeration from 2000 is known as CensusPlus. In this programme a sample will be selected (blocks or block clusters) for enumeration by a special method just after the census. Prior to the mailout of the census form interviewers called ICM-interviewers (integrated coverage measurement) will visit sample areas to list all the housing units to construct an independent listing of housing units and addresses, like that of the U.K. listing of all buildings. These lists will then be compared to the Master Address File (MAF) which is the main frame of enumeration and non-response follow up (NRFU). The housing units that were found by the ICM-interviewers but missed in the MAF and housing units that were included in the MAF will be followed up in the coverage estimates.

Just after the census day ICM-interviewers will go into the field to check the completeness of the MAF. They will estimate the number of housing units that were omitted from the MAF as well as the number of people missed from the census count as they were not in the MAF and hence not delivered the census form. They will also establish the vacant housing units. ICM-interviewers will also visit the housing units which were in the MAF but not in the independent lists.

In the sample areas the housing units that were in the MAF will be followed up as their census form comes back. ICM-interviewers will visit households for a computer-assisted personal interview only if they failed to contact the respondent for a computer-assisted telephone interview. Information will be collected in two parts from each of the respondent households. First, in order to construct a roster of persons living in the household, the interviewers will collect similar information but there may be more detail than on the census form. Some probing questions may also be included in the first part of the interview. The interviewers then disclose the original roster of the census form from the computer to check the discrepancies. In the second phase interviewers will attempt to resolve these discrepancies, if any, in order to come up with an accurate roster.

The vacant households established both by the ICM-interviewers and NRFU-interviewers will be revisited by the ICM-interviewers in order to verify that the units are in fact vacant, or when a unit is not vacant, to conduct an interview.

The information obtained from interview and reinterview will be gathered together to calculate the final estimate of the population total and this will be believed to be more accurate than the census counts or any other post enumeration survey.

The main advantage of CensusPlus is that the operation of the programme will be started just after the census day and hence the estimate will not be affected by memory error. Second, independent lists of housing units in the sample areas may help to identify many missing housing units and missing people from these housing units, which will definitely improve the coverage of the population. Third, a follow up interview of the MAF housing units in the sample areas who returned the census form will help to identify erroneous enumeration. Finally, the programme may produce the final report more quickly than the time required by the present post enumeration program.

A disadvantage of the method is that the ICM interview will not be completely independent of the census, because names from the previous response will be available for matching and reconciliation on the spot. Correlation bias may be present in the CensusPlus estimates because the same people who are missed by the census may also tend to be missed by the CensusPlus methods. As some of the ICM operation will start just after the census day, it may overlap in time with the census operation. For example ICM-interviewers may, by mistake, visit some housing units who did not yet returned their NRFU form. This may effect ICM and census counts in either direction.

The method is not so far tried except in the 1950 Post Enumeration Survey, in which the U.S. Census Bureau measured coverage error by repetition of census enumeration methods, in a more thorough and refined form, on a sample basis (U.S. Bureau of the Census, 1960). The U.S. Census Bureau has decided to test a version of the CensusPlus method in the 1995 census test in order facilitate identification of residence on census day and to improve the ability to produce final census results by legal reporting time. The procedure has also been designed to distinguish housing unit coverage errors in the Master Address File (MAF) from coverage errors that occur during census enumeration of housing units (Steffey et al., 1994).

2.8 SuperCensus

The methodology of the SuperCensus is almost same as that of CensusPlus. That is, like CensusPlus, SuperCensus selects a sample of blocks or ED workload and conducts the enumeration with special methods similar to those described for used in CensusPlus. The population estimates are based on applying the ratio of people to housing units observed in the sample blocks or ED workload to the total number of housing units.

The advantage of the SuperCensus is that it can be conducted simultaneously with the census. A disadvantage is that people missed by the regular census enumeration methods may also tend to be missed by the SuperCensus methods (Mulry, 1994). The variance of the population estimates tend to high because the estimation cannot use ratios to the census results to reduce variance, but must rely instead on crude preliminary measures of size, such as prelist housing unit counts, to reduce variance (Wolter, 1986).

2.9 Conclusions

In comparing the above methods it is apparent that all the methods have some advantages and also some disadvantages. However, which method a particular country will use for her coverage evaluation of the census does not depend only on the advantages of the method but also depends on the existence of facilities and resources of the country. To my knowledge, in the U.K. there are no administrative records which cover the whole population and at present ONS do not have any plans to use administrative records for census coverage purposes. The only possible ones are the Family Health Service Authority registers or the new Addresspoint but these are subject to error. If ONS plan to do so, they may create an administrative list by combining several lists and effort should be made to expand the content and improve the quality of the records over the next several years.

Demographic estimates are the most reliable estimates at the national level. Traditional methods for demographic analysis produce estimates of the national population cross-classified by age and sex. These estimates are generally used to check the coverage of the total population of the census counts. Demographic analysis does not produce estimates at local level or estimates by ethnic groups due to the lack of accurate information. However, OPCS is producing subnational demographic estimates from 1981 but these are believed to be affected due to inappropriate internal migration information.

The main methodological deficiency of the present post enumeration survey estimates of the U.K. is that they do not estimate people missed by both the census and survey. Moreover, the method is unable to estimate undercount rate by age, sex, and race. It is also unable to estimate local area populations. Considering all these facts, we believe use of dual-system method can give a better solution under the present circumstances and suggest that the dual-system method of estimation be used in the 2001 census evaluation programme.

Chapter 3
1991 Census and Census Validation Survey (U.K.)

3.1 Census

The 1991 decennial census of the population of the U.K. was taken on 21st April. In order to strengthen the field organisation the country was divided into 2,500 Census Districts (CD). These Census Districts were again divided into 130,000 small parts, known as Enumeration Districts (ED), each containing on average 200 household spaces (private households). The enumeration process also covered communal establishments; or other non-private accommodation such as hostels, boarding houses, prisons and defence establishments.

To minimize the error in the census, a three-tier system of Census Managers, Census Officers and Enumerators was introduced. The Managers were responsible for areas of the country each with a population of about half a million persons. They were also responsible for the recruitment and training of the Census Officers. The Census Officers in turn were responsible for the recruitment and training of the Assistant Census Officers and Enumerators. The main task of the Assistant Census Officers was to control the quality of the enumerators' work and to carry out checking duties on behalf of the Census Officers.

The main task of the census enumerator is to list in their Enumeration Record Book (ERB) the addresses of all buildings, during an advance round of their ED(s), indicating whether they were private residences, partially residential or non-residential premises, and to take care to record separate entries for each household space found in multi-occupied buildings.

The enumerators' other main duties were to enumerate every household present on census night and every person who spent census night in a communal establishment or other non-private accommodation, within their ED(s). They were responsible for the delivery and collection of census forms.

Enumerators issued a census form at each listed address (except where it was known that no one would be present on census night) for completion by each householder and identified as far as possible, the occupants and the number of households. For the households where it

29

was known that no one would be present on census night, no census form was required to be completed for that household address. Such addresses were recorded as "household absent" and the enumerators left the census questionnaire at these addresses, along with a leaflet inviting the absent residents to complete the questionnaire and post it back to the census offices when they returned home.

During the collection of forms, the enumerators were asked to check that the form had been properly completed, that no persons or any living accommodation had been missed; and that the occupancy status of vacant accommodation or absent households (or recorded at the delivery stage) had been correct. After collecting the form, enumerators were also asked to check them more thoroughly at home. It was expected that enumerators would contact the householders concerned, if any were found to be incomplete and obtain the missing information.

Finally, for each household, the enumerators entered in the ERB, the number of persons recorded on the form as present and those recorded as usually resident but absent, and the total of these two numbers. These figures were then aggregated, to form an ED summary, which in due course, was collated with others to produced preliminary population counts of persons (and households) up to the national level.

3.2 Census Validation Survey (CVS)

The main objectives of the census validation survey were to evaluate the quality and coverage of the 1991 census. More clearly, the objectives were:

i)   to check whether all eligible persons were enumerated correctly in the census;
ii) to check the extent of misclassification of unoccupied residential accommodation by census enumerators; and
iii) to assess the quality of answers given to census questions.

3.3 Sample Design and Methodology of (CVS)

To check the enumeration error, that is, to measure the quality and coverage of the 1991 Decennial Census, the U.K. Office of the Population Censuses and Surveys (OPCS), carried out a survey, known as Census Validation Survey (CVS) around six weeks after the Census by professional interviewers from the government field force. The CVS enumerators carried out the survey in six stages which were

1. a lisiting of all addresses in the sample Enumeration District (ED) to assess the coverage of buildings by comparing with the listing of the census enumerators
2. a check of buildings listed as non-residential
3. a check of addresses listed as absent

4. a check of addresses listed as vacant

5. a check of multioccupied addresses and

6. a reinterview of about five households in each enumeration workload

3.3.1 Sample Design:

For the CVS the sample was drawn in three stages. In the first stage, out of 2,500 CDs a sample of 300 CDs (13 per cent of the total number of CDs) were selected for Great Britain with probability sampling based on the 1981 Post Enumeration Survey (PES) experience together with the prevailing time and cost consideration (Heady et al, 1994). Among the 300 CDs, 270 CDs were allocated to England and Wales and 30 to Scotland. It seems to me that the sample size for Scotland is relatively small which may affect the undercount estimates. The situation will be worsen if the actual response rate were lower than the expected rate.

Table 3.1: Distribution of Sampled Census Districts

Strata	Number of Census District samples
Inner London	29
Formal Metropolitan and mixed metropolitan	109
Non-metropolitan	132
England and Wales	270
Glasgow	08
Rest of Scotland	22
Scotland	30
All	300

Source: OPCS (1994)

At the second stage, EDs were arranged (mostly within wards) in blocks of normally four adjacent EDs, or rather ED workloads, within each of the selected CDs. These blocks were constructed using maps and an appropriate route for covering the census districts. In some cases several EDs were grouped together to form an 'enumerator workload' because they contained only a small number of households. Some EDs with large communal establishments expected to have more than 100 people were designated as special EDs and were excluded from the CVS sample. In England and Wales all other EDs were graded according to the expected difficulty of carrying out the enumeration. An ED which was assumed easy to enumerate was graded 'A' through to grade 'G' the most difficult one. In Scotland areas were graded according to whether they were wholly urban, urban and rural mixed, or wholly rural. After completing the construction of blocks, each block was selected from each selected CD with probability proportional to the expected number of households they contained, weighted to take account of the expected difficulty of enumeration in the area. The expected numbers of households were the estimates used when planning the

31

census. The measurements of expected enumeration difficulty were also derived from the area classifications used for census planning. The EDs which were graded as difficult were over -sampled in the CVS because it was felt that they were more likely to produce higher-than-average error rates.

The country was divided into distinct major strata for sampling purposes, and within each major stratum further implicit stratification was achieved by means of the order in which CDs were listed. Within each major stratum sampling was done using a fixed selection interval--defined in terms of weighted population of households-and a random start.

3.3.2 Stratification

The CDs for England and Wales were grouped into three strata: Inner London, former metropolitan and mixed metropolitan, and non-metropolitan areas. The 270 CDs for England and Wales were divided between strata in proportion to their total weighted population: 29 were sampled in the `Inner London' stratum, 109 in the `former metropolitan and mixed metropolitan' stratum, and 132 in the `non-metropolitan' stratum (Table 3.1). The same criteria were used to allocate CDs in Scotland to two major strata: Glasgow in one and the rest of Scotland in the other.

In the third stage, from each selected block, interviewers selected different subsamples. On the average approximately 20 households were selected. Hence from 300 blocks, finally 6000 households were selected for quality check.

3.3.3 Selecting the Sample to Check Enumeration Error:

Visual List (VL)

Like the census enumerators, the CVS interviewers' first task was to list, from observation, the addresses of all building/building units within the boundaries of the selected EDs (workloads). These Visual Lists were independent of the census list, in order to avoid enumerators' mistakes influencing the interviewers when they come to make their listing.

The CVS interviewers compared their list against the ERB in order to identify additional building units i.e. those units which census enumerators have missed and addresses incorrectly included by the enumerators, although they were not within the allocated workload. Both of these discrepancies are potentially serious problems for the census. The first can lead to undercounting and the second can lead to overcounting.

Non-Residential Premises

The CVS interviewers renumbered the non-residential premises, which were distinguished by the entry 'NR' in the ERB by the census enumerators and by the absence of a form. From this new list interviewers selected every second non-residential address. The selected addresses were revisited by the CVS interviewers to check if they had actually been non-residential on census night.

Vacant Sample

From ERBs, the CVS interviewers renumbered consecutively those addresses which had been enumerated as vacant households by the census enumerators on census night. The vacant accommodation included new, never occupied property, property under improvement, second residence, holiday accommodation and students' accommodation. After renumbering, interviewers selected 1 in 2 samples, starting with a pre-specified random number. From the selected EDs, the resulting sample sizes were 6,790 vacant household spaces. The CVS interviewers re-visited all the selected addresses to ensure that they had in fact been vacant at the time of the census; an interview conducted with an occupier, if any; failing that with a neighbour or anyone else who was able to confirm the occupancy status on census night.

Absent Sample

The CVS interviewers visited all the spaces recorded as absent by the census enumerators in their workload areas to ensure that the classifications were correct and also to obtain information about the household members if any.

Multi-Household Sample

In the selected quality check households, if any address was found as a part of the Multi-household address (a house or purpose built flat--contained more than one household space), CVS interviewers were asked to record that address. From the ERB, details of up to six Multi-household addresses in each sampled CD were taken for reinterview. The CVS interviewers conducted the interview at any households which the enumerator had missed.

Quality Check Sample

To check the quality of the census enumeration a sample of 5,991 households were selected from all private households who returned a census form within the sampled EDs. Like the Vacant/Absent sample CVS interviewers renumbered all the eligible households to construct the sampling frame for the quality check sample and draw the sample, starting with a number pre-specified by the Social Survey Sampling Branch of OPCS. The CVS interviewers interviewed all the selected households in order to assess the quality of the answers given in the census. There were three interview schedules for the quality check sample, relating to housing, household composition and individual characteristics.

3.3.4 Handling unresolved addresses

When the CVS interviewers drew their samples, the field operations on the census had not quite finished. As a result, the Enumerator's Record Books would not contain the final census classification of each address, that is, on one hand the ERB contains some of the unresolved addresses and on the other hand some of the classifications changed between the time the CVS sample was drawn and the time the data were eventually entered in the census computer. In the case of unresolved addresses the CVS interviewers selected up to eight unresolved addresses from each selected CD and treated them in the same way as that of census-absents and census-vacants addresses. The late census reclassifications were handled during computer matching.

3.4 CVS Results and Discussion

The 1991 Census recorded 49,890,273 people as resident in England and Wales (excluding visitors but including residents who were recorded as absent on Census night). Among the recorded people 47,055,200 were recorded in England and 2,835,073 in Wales. The resident population in Scotland was 4,998,567 (OPCS, 1992).

The initial estimates of the 1991 Census Validation Survey identified a net undercount of 272,000 persons in Great Britain. Table 3.2 gives the overall estimates, that is estimates from all the six samples. According to the CVS's preliminary estimates the total net under enumeration was divided into three categories Resident, Visitor and Not-known of which the net under enumeration of the Residents was 143,000, Visitors were 96,000 and the Not-known was 32,000. The estimates take account both of the straightforward under and overenumeration, and also of misclassification (e.g. a self-contained space having wrongly been classified as non-self contained).

Table 3.2: Preliminary CVS Estimates for Great Britain (in thousand)

Persons	Census	CVS	Difference Between Census and CVS
Resident	54,426	54,569	-143
Visitor	1,001	1097	-96
Not-known	111	143	-32
Total	55,537	55,809	-272

Source: OPCS (1994)

The results from the CVS are not the best estimates that can be derived from the survey because they do not take account of any of the problems which were encountered in the CVS design itself. The problems were identified and described in detail in Chapter 3 of the CVS Coverage Report (OPCS, 1994). Allowing for some of these problems (such as, uncertainties about ED boundaries, occupied accommodation misclassified as vacant or non-residential by enumerators, overestimation of undercoverage at census-vacant and census-non- residential address etc.) the best CVS estimate is given in Table 3.3. The survey's estimate of the net underenumeration in the census came to 288,000 individuals, of which 162,000 were residents, 98,000 were visitors and 28,000 were people whose residential status was not known.

However, this best estimate of the population is much less than the demographic estimate of the population which is 56,207,000 for Great Britain, that is, approximately 1,300,000 more than the Census count for the resident population. This figure reduced to 1,200,000 after adjusting for definitional differences.

Table 3.3: Best CVS Estimates for Great Britain (in thousand)

Persons	Census	CVS	Difference Between Census and CVS
Resident	54,377	54,539	-162
Visitor	997	1095	-98
Not-known	113	141	-28
Total	54,486	55,774	-288

Source: OPCS (1994)(The Census figures in the Table 3.3 differ from those given in Table 3.2 because the present table excludes the Wrongly Included samples and those households in the CV samples that said they had returned a census form).

One popular way to check the accuracy of the CVS estimate is to compare the adjusted census count with the demographic estimates. The demographic estimate and the adjusted census count should be in close agreement. However, as we have already mentioned, for 1991, they were not. The overall difference for England and Wales was 717,000.

By comparing the adjusted census count with that of the estimated population as shown in Figure 3.1 we can analyse the difference by age and sex. The important features of Figure 3.1 are:

(a) The adjusted Census figures are substantially below the estimated figures for people aged under 45.

(b) Proportionately, the discrepancy is greatest for males and younger adults. The difference exceeded 2.5 per cent from age 19 to age 31, and exceeded five per cent from age 21 to 31, with a maximum of nearly six per cent at age 27.

(c) For the population between the ages of 45 and 80 the Census counts are similar to the demographic estimates. Here the differences are mostly one per cent and often less than 0.5 per cent.

(d) After age 70 the differences increased modestly. Over the age of 80 the estimated figures are higher. For men aged 85 and over, the adjusted census count was nearly nine per cent below the population estimate.

(e) The differences for the boys under 10 are not so large while the differences for teenage boys aged 14-17 are negligible.

(f) For the female population the differences are in the same direction as those of the males but smaller in size. For example the maximum difference, at age 27, is about 2.5 per cent, compared with six per cent for men.

Comparison by sex ratio

Figure 3.2 shows that the ratio of males: females in the demographic estimates is about 1.05 at age zero. This ratio remains fairly constant until age 20; then decreases to level off at about 1.00 from the mid-thirties to mid-fifties; with a sharp decrease thereafter at the older ages. In contrast, the census count and the adjusted count also both start about 1.05 but at age 18 or 19 the ratio drops rapidly to reach a minimum in the mid-twenties of 0.97 for the census count and 0.98 for the adjusted count. The ratio then recovers to reach much the same value as that for the demographic estimates for people in their late forties. The adjusted census ratio, thereafter, sometimes exceeds the demographic estimates.

Table 3.4: Adjustments to the 1991 Census

1991 Census count of usual residents for England and Wales	49,890
Timing difference: changes between census day (21 April) and mid-1991	+43
Definitional difference: net student balance	+58
Allowance for underenumeration in census over-imputation in processing Net underenumeration arising from missed/misclassified dwellings Omission of individuals in responding households Under enumeration of children under 1 Under enumeration of Armed Forces and their dependents	-85 +200 +177 +21 +70
Adjusted census count	50,383

Source: OPCS (1992). E and W means England and Wales.

The OPCS experts believed that there may be four possible reasons why the sex-ratios and the census count, the adjusted census count and the estimated populations were different. These four possible reasons are:

(1) errors in the baseline figures of the demographic estimates provided by the 1981 census.
(2) errors in rolling the estimates forward from the 1981 base.
(3) errors in adjusting the 1991 Census count and
(4) errors in the 1991 Census.

Table 3.5: Difference Between A.C.C. and R.F.E. for Mid 1991 (in thousands)

	Males	Females	Persons
Adjusted census count	24,482	25,902	50383
Rolled-forward estimates	24,892	26,063	50,955
Difference	410	161	572

Source: Population trends no. 71, 1993. A.C.C means adjusted census count and R.F.E means rolled-forward estimates.

By examining all the above four causes carefully, the OPCS experts reached the conclusion that most of the errors were probably attributed to the 1991 Census. Though the methodology of the 1981 PES was very similar to that of the 1991 CVS, comparison between adjusted census count and rolled-forward estimates showed more difference in 1991 (572,000 more than census count) than that of 1981 (108,000 more than rolled-forward estimates). Moreover, the age structure pattern of 1981 was very different than that of Figure 3.1. So, it is reasonable to believe that some other factors may be responsible for the different pattern of bias in 1981 compared to 1991.

In roll-forward estimates, the previous year's resident population is aged-on by one year. After ageing, the annual estimates are updated by adding on births, deducting deaths, and allowing for net migration balance. As the births and deaths registration system in the U.K. is believed to be almost complete, roll-forward estimates could be affected only by migration factors. After a review of the International Passenger Survey (IPS), "it was felt that IPS figures were more likely to understate than overstate immigration"(Heady et al, 1994). Thus we can conclude that, only a small part of the differences would be due to errors in the migration components.

The reasons for considering that errors in the 1991 census were the main cause of difference, is that the 1991 census was carried out in the traditional way without taking into account changes in society over the last ten years. For example, "there are more one-person households; more people are unwilling to answer the door to strangers; there is more homelessness; and there had been concerns that the unpopularity of the community charge might affect the census, even though the two exercise were not connected in any way" (Population Trends No. 71, 1993).

3.5 Conclusions

The overall conclusion of accepting demographic estimates is based on the fact that there was no reason to think that figures produced by demographic methods were overestimates. Furthermore, from the above explanations it is likely that the CVS underestimates the 1991 census undercount. When compared with the demographic estimates it was observed that most of the people missed in the 1991 Census were of the age groups 20 to 35. Past experience (Britton and Birch, 1985) had shown that people in their twenties are more difficult to enumerate: men more so than woman. This is precisely the pattern found in the 1991 census. Moreover, this pattern of age and sex differences does not fit any of the other explanations.

Table 3.6: Mid-1991 Population Estimate for E and W (in thousands)

1991 Census count of usual residents for England and Wales	49,890
Timing difference: changes between census day (21 April) and mid-1991	+43
Definitional difference: net student balance	+58
Visitors enumerated but not recorded as usual resident	+200
Over-imputation in processing	-115
Under-enumeration of special groups Children under 1 Armed Forces and their dependents Over 80 years old	+21 +42 +63
Under-enumeration (net) identified in CVS arising from missed/misclassified dwellings Omission of individuals in responding households	+178 +177
Other unexplained under-enumeration balance	+547
Final rebased mid-1991 population estimate for England and Wales	51,100

*Source: OPCS (1993)

Chapter 4

Comparison Among Different Dual System Estimates

4.1 Introduction

Dual record systems are employed in a number of countries to estimate the undercount/overcount rate of the census by using the C-D technique. For the application of the C-D technique, the first record is the census and the second record is the sample count and it is assumed that the two data collection systems are independent in the sense that neither has access to the other's results and a third independent check is made on events picked up by either system to verify that they are correct. Normally, events picked up jointly are assumed to be correct. Sample data are matched with census data, on a case by case basis. The outcome can be distributed in the form of a 2x2 contingency table (Table 4.A, same as Table 2.1), where x_{22} designates an unknown number of events not observed by either method, and therefore not available from any tally of the survey results.

Table 4.A. Outcome of the Matching Process

Census	Sample (PES)		
	In	Out	Total
In	$x_{11} = M$	x_{12}	x_{1+}
Out	x_{21}	$x_{22} = z$	$x_{21} + x_{22}$
	$x_{+1} = \check{N}_p$		$x_{++} = \check{N}_p$

The proposed estimator of x_{22} by Chandra-Sekar and Deming (1949) is:

$$x_{22} = x_{12}x_{21}/x_{11} \tag{4.1}$$

Hence the total estimated number of persons, $\check{N}_T$ is

$$\check{N}_T = x_{11} + x_{12} + x_{21} + x_{22} \; [= x_{12}x_{21}/x_{11}] \qquad (4.2)$$

Chandra-Sekar and Deming assert that when the chance of an event being missed by the first system is independent of the chance of the same event being missed by the second system, equation (4.1) is a consistent estimate of x_{22}.

In reality the assumption of independent collection is unacceptable. Jabine and Bershad (1968), Greenfield (1975), El-Sayed Nour (1982), and many others argue that, in particular where the source of data is a human population, there are many possible reasons for which data can be missed systematically by both methods of data collection.

The correlation between the two methods is,

$$r_x = (x_{11}x_{22} - x_{12}x_{21})/[(x_{11} + x_{12})(x_{11} + x_{21})(x_{12} + x_{22})(x_{21} + x_{22})]^{1/2} \qquad (4.3)$$

Based on real observations, Chandra-Sekar and Deming (1949), Jabine and Bershad (1968) and Greenfield (1975), were convinced that the association between these two collection systems is positive. When positive correlation between the two recording systems exists, this means that x_{11} in equation (4:1) will be relatively higher than would be the case if the systems were independent and hence x_{22} will be an underestimate. In other words, when the second system is more likely to pick up events recorded by the first system than events missed by the first system, positive correlation will arise. The reverse applies for negative correlation (Greenfield, 1976). Chandra-Sekar and Deming, therefore, suggested that their method will give better results if applied to homogeneous sub-groups of the data and that the total estimate be obtained by building up from these sub-groups. The underlying argument is that if the association for each sub-group is near zero, while the association for all sub-groups combined is not zero, then a less biased estimate of x_{22} will result.

4.1.1 Greenfield Method

Greenfield (1975) argues that while the method of sub-grouping offers an improved estimate, it still suffers from the defect that independence within sub-groups is assumed. He therefore proposed that the C-D estimate of the number of missed events should be regarded as a minimum estimate. His proposed estimator for the lower limit to the value of x_{22} is the same as that of Chandra-Sekar and Deming (1949) which is

$$x_{22} = x_{12}x_{21}/x_{11}$$

An upper limit to the value of x_{22} is derived by taking r_x as Pearson's correlation co-efficient and then writing the equation in the quadratic form for x_{22} and solving for x_{22} as

$$x_{22} = -1/2B + (A + B^2/4)^{1/2} \tag{4.4}$$

where

1. $A = (x_{12}x_{21})[x_{12}x_{21} - r_x(x_{11} + x_{21})(x_{11} + x_{12})]/ r_x^2(x_{11} + x_{21})(x_{11} + x_{12}) - x_{22}^2$

$$\tag{4.5}$$

2. $B = [r_x^2(x_{12} + x_{21})(x_{11} + x_{21})(x_{11} + x_{12}) + 2x_{11}x_{12}x_{21}]/ r_x^2(x_{11} + x_{21})(x_{11} + x_{12}) x_{11}^2$

$$\tag{4.6}$$

3. $r_x = 1/2[r_{x(max)} + r_{x(min)}]$ $\tag{4.7}$

4. $r_{x(max)} = x_{11}/(x_{11} + x_{21})(x_{11} + x_{12})^{1/2}$ $\tag{4.8}$

5. $r_{x(min)} = - [x_{12}x_{21}/(x_{11} + x_{21})(x_{11} + x_{12})]^{1/2}$ $\tag{4.9}$

The technical details of the estimation procedure are given in Appendix 1.

4.1.2 El-Sayed Nour Method

El-Sayed Nour (1982), also did not agree with the suggestion of dividing the data into homogeneous sub-groups and argued that a better way of dealing with the association bias in estimating x_{22} is to make assumptions concerning the value of the association index r where

$$r = A(x_{11}x_{22} - x_{12}x_{21}) \tag{4.10}$$

where A is an appropriate positive constant. He presents an alternative approach to the estimation of x_{22} which preserves the main characteristics of the C-D technique, but takes into account the lack of independence between the results of the two collection procedures.

He defined the properties of the C-D technique in the context of demographic application, and by taking those properties into account he derived the upper and lower limit of x_{22} and suggested the lower limit as an estimator for x_{22}. His given estimator is

$$x_{22} = 2x_{11}x_{12}x_{21}/(x_{11}^2 + x_{12}x_{21}) \tag{4.11}$$

The technical details of the estimation procedure are given in Appendix 2.

On the other hand, in the CVS investigation to check the quality and coverage of the 1991 census, they have six different samples from the same ED blocks of which only the CVS listing of housing units is independent of the census listing. That is, among the six samples, five are related to the census enumeration. Under this situation if we apply the C-D technique directly for x_{22} we may have a less precise estimate of missing events. Therefore, it is clear that if we estimate the missing events by applying all the above three methods, that is, C-D, Greenfield, and Nour methods we will be able to observe the whole range of missing events, which in turn will help us to draw conclusions. Therefore, in the following we will describe the estimation procedures of the above mentioned three methods and then compare the estimates of the missing values.

4.2 National Estimate of Net Undercount/ Overcount:

To estimate the net undercount/overcount at national level, we consider the nine regions of England and Wales as defined in the 1991 census enumeration. We assume that each of the regions is divided into some CDs and each CD is again divided into a number of ED blocks. Each ED block gives an independent estimate of the total number of people enumerated by the 1991 census. This estimate will be obtained in two stages. First the C-D estimate will be computed for the total population of each ED block.

Second, this estimated total population of each ED block will be divided by the probability of selection of the CDs. This will give an estimate of the total population of the stratum which is

$$\hat{Y}_{ppsi} = M_i(AE)_{ij} \, / \, m_i P_i \tag{4.12}$$

where =

$\hat{Y}_{ppsi}$ is the estimated total population of the region from the ith CD

M_i is the number of ED blocks in the ith selected CD

m_i is the number of ED blocks selected in the sample from the ith CD

P_i is the probability of selection of the ith CD which is the first sample unit

$(AE)_{ij} = N^*_{Tij}$ is the adjusted estimate of the total population in the jth selected ED blocks of the ith CD; this could be obtained from the C-D method, the Greenfield method, or the Nour method.

An unbiased estimate of the total population of the stratum is given by

$$\hat{Y}_{PPSi} = 1/n \Sigma^n_i \, \hat{Y}_{PPSi} \tag{4.13}$$

where n is the number of selected CDs

The total population of the whole country will be estimated in the following way:

$$\check{N}_T = 1/k\Sigma_1^k \, [\hat{Y}_{PPSi} / x]X \qquad\qquad (4.14)$$

where

K is the number of regions
x is the census enumerated region total and
X is the census enumerated total population of the country.
$\hat{Y}_{PPSi}$ is the estimated total population of the ith region

The estimate of net undercount/overcount rate is therefore

$$\check{R}_n = 100(1 - N_c / \check{N}_T) \qquad\qquad (4.15)$$

$\check{R}_n$ is the undercount/overcount rate
N_c is the enumerated total population by the census
$\check{N}_T$ is the estimated total population

4.3 Data for the Dual System Method

To estimate the population of England and Wales, let us consider that the country is divided into nine regions and the 1991 census enumerated population of these nine regions is 49,193,915. We assume that each region is divided into a number of Census Districts (CD) and only one CD from each region is selected by probability proportional to size. To draw the CD by probability proportional to size, generally last census year's population of the CD is considered. Each CD is again divided into Enumeration Districts (ED). These EDs are then arranged in blocks, taking four adjacent EDs in each block. To estimate the total population of the country, the design of the sample was two-stage with CDs as first stage units and ED blocks within the CD as second stage units. CDs were selected with replacement and with probability proportional to the census population as recorded in the last census, whereas ED blocks in a CD were selected with equal probability. The data are given in Table 4.B and Table 4.C. The figures in column 2 of Table 4.B are the 1991 census enumerated populations of the regions. All other figures of Table 4.B and Table 4.C are hypothetical.

Table 4.B. Enumerated Population of Regions

Region	Total persons (1991)	Selected CD	CD Population (Hypothetical)
Eng + Wales	49193915		
North	3018679	$CD_{1.4}$	750
York + Humb	4796,562	$CD_{2.3}$	854
East Midland	3919483	$CD_{3.5}$	998
East Anglia	2018617	$CD_{4.2}$	823
South East	16793683	$CD_{5.7}$	794
South West	4599685	$CD_{6.4}$	862
West Midland	5088565	$CD_{7.2}$	924
North West	6146776	$CD_{8.6}$	768
Wales	2811865	$CD_{9.1}$	820

Here, for example, $CD_{1.4}$ means, selected CD number 4 of region 1 (the north)

Table 4.C. Hypothetical Population of ED Blocks

Region	Selection Prob. Of the CD (p_i)	No of ED in CD (M_i)	Population in the selected ED blocks				
			EDs'	T_p	B_p	M_p	F_p
North	.000248	11	$ED_{14.9}$	64	01	31	33
York + Humb	.000282	14	$ED_{23.3}$	56	04	29	27
East Midland	.000254	12	$ED_{35.2}$	87	07	44	44
East Anglia	.000407	11	$ED_{42.5}$	76	04	40	36
South East	.000047	12	$ED_{57.7}$	69	00	35	34
South West	.000187	10	$ED_{64.1}$	80	03	42	38
West Midland	.000181	11	$ED_{72.4}$	86	04	44	42
North West	.000124	14	$ED_{86.5}$	59	01	30	29
Wales	.000291	13	$ED_{91.3}$	62	05	32	30

Here $ED_{14.9}$ means selected ED number 9 of CD 4 of region 1.

Column 2 represents the selection probability of the CD from the region. For example, the probability of row one of Table 4.C is .000248 and is calculated by dividing 750/3018679 where 750 is the CD population (hypothetical) and 3018679 is the 1991 census counted total population of the North region.

Column 3 represents the total number of ED blocks in the selected CD (hypothetical).

T_p, represents the total counted population of the selected ED (hypothetical).

B_p, represents the total counted black population of the selected ED (hypothetical).

M_p, represents the total counted male population of the selected ED (hypothetical).

F_p, represents the total counted female population of the selected ED (hypothetical).

4.4 Procedure of the C-D Technique

Using information from all the six CVS samples and by giving appropriate weight for unequal probabilities of selection in each sample, we estimate the population total for each of the following cases by applying the C-D technique as defined before.

1. National estimate of undercount/overcount
2. National estimate by Race
3. National estimate by Sex
4. National estimate by Age
5. Estimate for Inner London
6. Estimate for Metropolitan areas
7. Estimate for Non-Metropolitan areas

In this section we calculate estimates (a), (b) and (c) using the C-D method.

By applying the C-D technique we can estimate the total population of these selected ED blocks separately as follows:

1. Step 1 Let us consider Enumeration District $ED_{14.9}$. Our problem is to estimate the total population of $ED_{14.9}$.
2. Step 2 Our model is (Table 4.1)

Table 4.1. C-D Model for Population Estimate

Census	CVS		
	In	Out	Total
In	$x_{11} = M$	x_{12}	$x_{cij} = N_{cij} - G - E - D - I$
Out	x_{21}	x_{22}	
	$\check{N}_{pij}$		N^*_{Tij}

where

$M = x_{11}$ is the population counted by the both methods, census and survey

x_{12} is the counted population in the census only

x_{21} is the population estimated from the CVS sample only

x_{22} is the population missed by the both procedure

N_{cij} is the enumerated population in the jth ED block of the ith CD in the census

x_{cij} is the corrected population in the jth ED block of the ith CD

$\check{N}_{pij}$ is the estimated weighted total in the jth ED block of the ith CD from the CVS sample

N_{Tij} is the estimated total population of the jth ED block of the ith CD by the C-D method of estimation

G is the number of persons incorrectly located geographically in the census

E is the number of persons incorrectly enumerated in the census (fabricated or not in sample)

D is the duplicate enumeration in the census

I is the number of persons who are enumerated in the census but have insufficient information for matching

Step 3 Estimate of $\check{N}_{pij}$

a) From the visual list, if any additional building is found which was missed by the census enumerator, a coverage check interview will be carried out. Say, for $ED_{1.3}$ we found three persons, who were not enumerated in the census count

b) From the quality check sample, let us draw a sample of size 5 from the 20 who were enumerated in the census. Let us also assume that during the quality check interview, we have found two persons who were not enumerated in the census. As we draw a sample of size 5 from 20, so our estimated total persons not numerated in the census count is $(20/5)2 = 8$. In the same way we will estimate the persons not enumerated in the census from the Non-residential household, Multi-household and Vacant/Absent household samples.

For estimating $\check{N}p_{1.3}$ let us consider that persons who were not enumerated in the census were found in the following ways:

a) 3 persons from the visual list sample
b) 8 persons from the quality check sample
c) 1 person from the absent household sample
d) 2 persons from the non-residential household sample
e) 2 persons from the multi-household sample
f) 0 persons from the vacant household sample

g) G+ E + D + I = 0 + 0 + 4 + 0 = 4, that is from the samples four persons were found counted twice in the census and hence must be subtract from the census count.

$\check{N}_{pij} = \check{N}p_{1.3} = x_{11} + (x_{21} = 3+8+1+2+2+0 = 16) = 58 + 16 = 74$

$x_{Cij} = x_{c1.3} = N_{c1.3} - G - E - D - I = 64 - 0 - 0 - 4 - 0 = 60$

Step 4 Estimate of $ED_{14.9}$ block's population by C-D Technique

At this stage of the estimation we counted or estimated the following:

N_{Cij} = 64 which is the census count

x_{Cij} = 60 which is the corrected census count

$\check{N}p_{ij}$ = 74 which is the estimated weighted total in the jth ED block of the ith CD from the CVS samples

Now from the census and the CVS sample's matching along with the above three pieces of information we can fit our model as follows: (Table 4.2)

Table 4.2. C-D Model for ED Population Estimate

Census	CVS		
	In	Out	Total
In	x_{11} = 58	x_{12} = 2	$x_{c1.3}$ = 60
Out	x_{21} = 16	x_{22} = .55	16.55
	$\check{N}_{P1.3}$ = 74	3	$N^{*}_{T1.3}$ = 76.55

where

M = x_{11} = 58 means that out of 60 corrected census count 58 persons were matched with the CVS count/estimate.

x_{12}= 2 means that out of 60 corrected census count 2 persons were not matched with the CVS count.

x_{21} = 16 means that out of 74 CVS estimated persons 16 were not matched with the census count.

x_{22} = $x_{12}x_{21}/x_{11}$ = 2x16/58 = .55

$\check{N}^{*}_{T1.3} = (\check{N}_{P1.3} x_{c1.3})/x_{11} = (74)(60)/58 = 76.55$ (4.16)

Step 5 The total population of the stratum is

$\hat{Y}_{ppsi} = M_i(CDE)_{ij}/(m_iP_i) = (11)(76.55)/.000248 = 3395363$ (4.17)

$M_i = 11$; $m_i = 01$; $P_i = .000248$

Following the same procedure we estimated the total population of all the regions from the selected ED blocks; the result is given in Table 4.8.

4.4.1 Estimate by Race

To estimate the undercount by race, post stratification technique will be applied. First, from the census count, total population for a particular race (say black) will be separated. In the second stage, from the CVS sample, the total black population will be estimated by the C-D method for each block.

For illustration let us consider the previous example. Total black population in the selected ED blocks = 1+4+7+4+0+3+4+1+5 = 36 (Total of column 6 of Table 4.C)

As the sample size is small let us consider all the selected ED blocks as a block, following the same procedure as before, we have:

a) 0 persons from the visual list sample
b) 4 persons from the quality check sample
c) 0 person from the absent household sample
d) 0 persons from the non-residential household sample
e) 0 persons from the multi-household sample
f) 0 persons from the vacant household sample
g) $G + E + D + I = 0 + 0 + 0 + 0 = 0$

$\check{N}_{PB} = x_{11} + (x_{21} = 0+4+0+0+0+0=4) = 30 + 4 = 34$ is the estimated black population in the combined blocks from the CVS samples.

Our model is (Table 4.3):

Table 4.3 C-D Model for Population Estimate by Race

Census	CVS		
	In	Out	Total
In	$x_{11} = 30$	$x_{12} = 6$	$x_{CB} = 36$
Out	$x_{21} = 04$	$x_{22} = .88$	4.88
	$\check{N}_{PB} = 34$	6.800	$N^*_{TB} = 40.80$

where

$x_{CB} = N_{CB} - G - E - D - I = 36-0-0-0-0 = 36$ is the corrected black population in the combined ED blocks.

N_{CB} is the enumerated black population in the combined ED blocks.

$\check{N}^*_{TB}$ is the estimated total black population of the combined ED blocks.

Calculations for the black population:

Table 4.4. Estimating Black Population.

CD	M_i	P_i	m_i	CDE	M_iCDE_i	M_iPi	M_iCDE/m_iP_i
CD_{1to9}	108	.00008	9	40.80	4406	.0007	6283133

where

M_i = 108 is the total ED blocks in the selected CDs.

P_i = total black population in the selected CD divided by total black population of the sample population = 460/5903270. Here both the figures are hypothetical.

m_i = 9 is the total number of selected ED blocks and

CDE = 40.80 is the dual system estimate.

Therefore, the estimate of net the undercount/overcount rate of the black population is:

$R_{nB} = 100(1 - N_{CB}/N^*_{TB}) = 100(1 - 5903270/6283133) = 6.04$ (overcount)

(4.18)

4.4.2 Estimate by Sex

Each of the selected ED blocks will be divided into two parts: male and female. Estimates for male and female, can then be estimated separately. For illustration let us consider the North region.

Total population of the North counted in the census = 3018679
Total male population of the North counted in the census = 1457472
Total male population of the selected CD (hypothetical) = 384
Hypothetical males in the selected ED is N_{CM} = 31

Estimate of male population from the selected ED by C-D Technique

Following the same procedure as before we have say:

a) 4 persons from the visual list sample
b) 8 persons from the quality check sample
c) 2 person from the absent household sample
d) 0 persons from the non-residential household sample
e) 0 persons from the multi-household sample
f) 0 persons from the vacant household sample
g) $G + E + D + I = 0 + 0 + 4 + 0 = 4$

$\check{N}_{PM} = x_{11} + (x^*_{21} = 4+8+2+0+0+0=14) = 25+14= 39$

$x_{CM} = N_{CM} - G-E-D-I = 31-0-0-4-0 = 27$

Our model is (Table 4.5):

Table 4.5. C-D Model for Population Estimate by Sex

Census	CVS		
	In	Out	Total
In	$x_{11} = 25$	$x_{12} = 26$	$x_{CB} = 27$
Out	$x_{21} = 14$	$x_{22} = 1.12$	15.12
	$\check{N}_{PM} = 39$	3.12	$N^*_{TM} = 42.12$

where

$x_{CM} = N_{CM1.3} - G - E - I - D$ is the corrected male population.

$\check{N}_{PM}$ is the estimated male population from the CVS samples.

N^*_{TM} is the estimated total male population for the ED block by the C-D method.

Therefore, the estimate of the total male population of the region is:

$$\check{N}_{TM} = 1/n\Sigma^n_i \, \hat{Y}_{PPSi} = M_i CDE/m_i P_i = 1748531 \qquad (4.19)$$

where,

$M_i = 11; \; m_i = 01; \; P_i = 384/1457472 = .000263$

Therefore, the estimate of the net undercount/overcount rate of the male population is:

$$R_{nM} = 100(1 - N_{CB} / N^*_{TM}) = 100(1 - 1457472/1748531) = 17.21(und) \tag{4.20}$$

Following the same procedure we will estimate the total male population of the stratum from all the selected ED blocks.

4.5 Procedure of the Greenfield Method

For purposes of comparison in this section we calculate the national estimate using the Greenfield method. The estimation procedure of the Greenfield method involves several steps.

Step 1: Estimation of N^*_t (primary estimated total population of the ED blocks).

In step one by utilizing all the six CVS samples along with census enumeration we will estimate the total population for each of the selected ED blocks. For example, let us consider the previous example of Table 4.2

where,

N_C = Census enumeration = 64

$G + E + D + I = 0 + 0 + 4 + 0 = 4$ (symbols are as before)

V_C = Weighted estimated population from the CVS samples who were not enumerated in the census = 16

Therefore,

$N^*_t = N_C - (G + E + D + I) = 64 - 4 + 16 = 76$

Step 2: Estimation of Z (initial estimate of population missed by both the methods)

Here we will match the census enumeration with the Visual Listing only. From the matching result, we will estimate Z from the following equation

$$Z = N^*_t - x_{11} - x_{12} - x_{21}$$

Table 4.6. Model for Estimating Z

Census	Sample	
	In	Out
In	$x_{11} = M = 60$	$x_{12} = 4$
Out	$x_{21} = 3$	Z

Therefore,

$Z = 76 - 60 - 4 - 3 = 09$

Step 3: Estimation of $\hat{r}_x$. (correlation co-efficient between census and survey data)

In this step we will estimate $\hat{r}_x$ from equation (4.7) by utilizing equations (4.8) and (4.9). For our present example

$r_{x(e)(max)} = (x_{11}Z - x_{12}x_{21})/ [(x_{11} + x_{12})(x_{11} + x_{21})(x_{12} + Z)(x_{21} + Z)]^{1/2}$

$\quad = [(60)(09) - (4)(3)]/ \sqrt{[(64)(63)(13)(12)]}$

$\quad = .66$

$r_{x(min)} = -[(x_{12}x_{21})/(x_{11} + x_{12})(x_{11} + x_{21})]^{1/2}$

$\quad = -[(4)(3)/(64)(63)]^{1/2}$

$\quad = -.05$

Therefore,

$\hat{r}_x = \frac{1}{2}[r_{x(max)} + r_{x(min)}] = .305$

Step 4: Estimation of x_{22} and N^*_T (final estimated total population of the ED blocks.

At this stage by matching the census data with the CVS samples case by case we will estimate x_{22} from equation (4.4) in the following way. Our model is (Table 4.7):

Table 4.7. C-D Model for Population Estimate

Census	Sample	
	In	Out
In	$x_{11} = 58$	$x_{12} = 02$
Out	$x_{21} = 16$	x_{22}

Now

$$A = (x_{12}x_{21})[x_{12}x_{21} - \hat{r}_x^2(x_{11} + x_{21})(x_{11} + x_{12})]/ \hat{r}_x^2(x_{11} + x_{21})(x_{11} + x_{12}) - x_{22}^2$$

$$= 32[32 - .93(60)(74)]/ [(.093)(60)(74) (58)^2]$$

$$= 4.13$$

$$B = [\check{r}_x^2(x_{12} + x_{21})(x_{11} + x_{21})(x_{11} + x_{12}) + 2x_{11}x_{12}x_{21}]/ \hat{r}_x^2(x_{11} + x_{21})(x_{11} + x_{12}) x_{11}^2$$

$$= [(.093)(18)(60)(74) + (2)(58)(2)(16)]/[(.093)(60)(74)-(58)^2]$$

$$= -3.776$$

Therefore,

$$x_{22} = -1/2B + (A + B^2/4)^{1/2}$$

$$= 4.65$$

And the total population of this ED blocks is
$N^*_T = 58 + 02 + 16 + 4.65 = 80.64$

Step 5 The total population of the stratum is

$$\hat{Y}_{PPSi} = M_i(AE)_{ij}/(m_iP_i) = (11)(80.64)/.000248 = 3576774 \qquad (4.21)$$

$M_i = 11; m_i = 01; P_i = .000248$

Step 6 Therefore, estimate of net undercount/overcount rate of the stratum population is:

$$R_{nT} = 100(1 - N_{CM}/N^*_{TM}) = 100(1 - 3018,679/3576774) = 15.60 \qquad (4.22)$$

Following the same procedure we will estimate the total population of the region from all the selected ED blocks.

4.6 Procedure of El-Sayed Nour Technique

Here we present the national estimate using the Nour method. We use the example of Table 4.2 from which the estimated value of x_{22} is

$$x_{22} = [2x_{11}x_{12}x_{21}]/[x^2_{11} + x_{12}x_{21}] = [(2)(58)(2)(16)]/[(58)^2 + (2)(16)]$$
$$= 1.09 \tag{4.23}$$

and the total population of the ED block is

$$\check{N}_T = x_{11} + x_{12} + x_{21} + x_{22} [=(2x_{11}x_{12}x_{21})/(x^2_{11} + x_{12}x_{21})] \tag{4.24}$$
$$= 58 + 02 + 16 + 1.09 = 77.09$$

Therefore, the total population of the stratum is

$$\hat{Y}_{PPSi} = M_i(CDE)_{ij}/(m_i P_i) = (11)(77.09)/.000248 = 3419315 \tag{4.26}$$
$$M_i = 11; m_i = 01; P_i = .000248$$

Therefore, estimate of net undercount/overcount rate of the stratum population is:

$$R_{nT} = 100(1 - N_{CM}/N^*_{TM}) = 100(1 - 3018,679/3419315) = 11.71(und) \tag{4.27}$$

Following the same procedure we will estimate the total population of the region from all the selected ED blocks.

4.7 Results and Discussion

In Table 4.8, the estimated total population of the strata as estimated by the C-D, El-Sayed Nour, and Greenfield methods are given.

Table 4.8. Estimate from C-D, Greenfield and Nour methods
(using hypothetical data from Table 4.B and 4.C).

Region	1991 Census Count	CD Method	El-Nour Method	Greenfield Method
North	3018679	3395363	3419315	3576774
York + Humb	4796562	4827640	4827640	4979438
East Midland	3919483	4179685	4201890	4194803
East Anglia	2018617	2171429	2180541	2213243
South East	16793683	17820078	17842373	18849153
South West	4599685	4451610	4464171	4472456
West Midland	5088565	5181739	5197348	5278900
North West	6146776	7007923	7015807	7012081
Wales	2811865	2963074	2977491	2971237

From Table 4.8 one can observe that the lowest and highest estimated values are given by the C-D and Greenfield methods respectively while the Nour method gives a value in between the two methods. Only in two cases, that is, in the region East Midland and North West the estimated values by El Sayed Nour are higher than the estimated values by Greenfield methods. This means that C-D assumption of r = 0, that is, the assumption of zero correlation understates the true correlation and hence results in an underestimate of the number of events missed by both methods of data collection. On the other hand the assumption of the Greenfield technique that the true correlation is located at the mid-point of its range overstates the correlation and hence results is an overestimate of the number of events missed.

In Table 4.9 the range of the undercount rate calculated from the estimated values of Table 4.8 are given. As can be seen, in case of the C-D method ranges from overestimate of 3.32 percent to underestimate of 12.29 percent, whilst in case of the Nour method ranges from overestimate of 3.04 percent to underestimate of 12.39 percent. The ranges is highest in case of the Greenfield method which is from overestimate of 2.84 percent to underestimate of 15.60 percent.

Table 4.9. Range of Estimated Undercount rate

Methods	Lower Value	Upper Value	Range
C-D Method	-3.32	12.29	15.61
Nour Method	-3.04	12.39	15.43
Greenfield Method	-2.84	15.60	18.44

4.8 Conclusion

The main objections to the use of a dual system are directed to the correlation bias and cost. Other problems of the dual system are matching bias, and particularly a high level of erroneous non-matches. A high rate of erroneous non-matches will result in a small value of x_{11}, and will overestimate the missing events. On the other hand, erroneous inclusions will result in a high value of x_{11}, and will underestimate the missing events. Actually, recent developments of computer technology and intensive research in this field help to improve the design as well as matching techniques of the dual system. The improving matching techniques of the dual system give us the opportunity to work fast and improve the completeness of coverage with reasonable cost.

The presence of correlation bias has almost same effect as that of erroneous non-matches. When correlation is positive the dual system underestimates missing events and when correlation is negative, overestimates missing events. To deal with this correlation bias Chandra-Sekar and Deming suggest classifying the events into homogeneous groups on the basis of age, sex, and other characteristics; estimating population separately for each groups; and summing to get the estimated total number of events. Jabine and Bershad (1968) give a hypothetical illustration of the appreciable reduction in bias achieved by making separate DSE estimates for strata and then summing. Unfortunately, examples of substantial gains from making separate estimates for population subgroups are few. The main difficulty with trying to minimize correlation bias by making separate estimates for subgroups is that it is very difficult to define subgroups which are homogeneous with respect to the probabilities of coverage. As a result, the correlation between sources within groups is not different from the overall correlation.

We therefore try to investigate how different methods of estimation improve the estimates of missing events rather than depending only on dividing the population into subgroups. The results of Table 4.8 give clear support for the assertion that the assumption of zero correlation of the C-D technique should be regarded as providing an underestimate of missed events and the Greenfield technique an overestimate. Estimation by the Nour technique gives values in between the C-D and Greenfield techniques, which implies that the estimates by the Nour technique are less affected by the correlation of the two methods of data collection. Moreover, the calculation of the Nour technique is very simple. Therefore, we believe wherever C-D technique is used, the extra calculation involved in applying the Greenfield and Nour technique in addition to that of the C-D technique will be useful.

Chapter 5
Triple System Estimation Using Log-linear Models

5.1 Introduction

In Chapter 2 we have discussed the dual system estimation methodology to estimate the total population including those missed by the census and PES surveys. We also discussed the correlation bias and some alternative methods of estimation to reduce this correlation bias. Another way of reducing correlation bias is to obtain additional sources of information on the population. By using a third independent source of information, the 2x2 table underlying the PES can be expanded into a 2x2x2 table in which only one of the 8 cells is unknown. Estimates of the unknown cell and thus of correlation bias and total population may be calculated under suitable assumptions. Zaslavsky and Wolfgang (1990); Darroch, J. N. et al (1993), discussed a number of methods to estimate the population of the unknown cell as well as the total population by using the three sources of information in log-linear models. In this chapter we focus on the following four models based on three sources of information:

1. Model of complete independence
2. Model of no second order interaction
3. Model of full quasi symmetry
4. Model of partial quasi symmetry

5.2 Data for Triple System Method:

For the purposes of population estimation we assume that the primary source of data is a census enumeration, the secondary source is a post enumeration survey and the third source is a further survey or administrative record list. Ericksen and Kadane (1985) called this list a "mega-list" which they suggested could be created by combining all the available sources to approximate a complete list of the population. Here, for our purposes, we considered three hypothetical lists of information. The first one is the census count, the second one is the estimated number of persons from the CVS and the third one is the address list (at present no such list is available in the U.K.; the ONS is developing a master address list by combining all the available administrative lists (such as NHS records, post office address lists etc.)).

We divided the population into four strata and deliberately chose the number of persons in each list differently from the others in the different strata to see how well the models fit the data in different situations. In stratum 1 we assume that the census enumerated population and the counted population from the address lists are similar but less than the number of people counted by the CVS interviewers.

In stratum 2 we considered that the census and the CVS count of population are similar but less than the number of people listed in the address lists.

In stratum 3 we considered that the census and the address list populations are similar but less than the CVS enumeration.

In stratum 4 we try to observe how well the models fit when CVS and address list population counts are similar but considerably less than census counts. We have five estimates for each stratum. The data are given below (Table 5.1): Column 1-3 of Table 5.1 are the census, the CVS and the address list estimated population respectively while column 4-11 represents the matching results of the above three sources of information.

Table 5.1. Hypothetical Three-source Data

CEN	CVS	ADD	x_{111}	x_{121}	x_{211}	x_{221}	x_{112}	x_{122}	x_{212}	x_{222}
Stratum 1										
225	240	227	201	08	09	09	07	09	23	--
202	214	205	176	08	12	09	10	08	16	--
226	235	228	200	08	11	09	10	08	14	--
227	238	231	201	09	11	10	08	09	18	--
244	258	248	218	10	13	07	12	08	15	--
Stratum 2										
247	246	274	208	13	14	39	15	11	09	--
228	224	251	189	09	10	43	14	16	11	--
218	221	235	177	11	12	35	13	17	19	--
205	211	218	159	12	10	37	18	16	24	--
207	217	229	168	13	14	34	16	10	19	--
Stratum 3										
222	263	220	190	11	09	10	08	11	56	--
234	277	230	193	12	11	14	19	10	54	--
245	284	241	201	16	12	12	11	17	60	--
248	288	242	198	12	14	18	17	21	59	--
239	286	232	195	12	11	14	22	10	58	--
Stratum 4										
222	263	220	190	11	09	10	08	11	56	--
234	277	230	193	12	11	14	19	10	54	--
245	284	241	201	16	12	12	11	17	60	--
248	288	242	198	12	14	18	17	21	59	--
239	286	232	195	12	11	14	22	10	58	--

5.3 Log-Linear Models

Estimating the counts of individuals using log-linear models is now accepted as one of the most useful methods. The use of three or more lists with a multinomial sampling model was first explored by Darroch (1958) for independent lists and was extended with log linear models to allow dependence among the lists by Fienberg (1972). Such models have been studied extensively by Goodman (1968), Bishop and Fienberg (1969), Mantel (1970), Fienberg (1972), Haberman (1974), Bishop, Fienberg, and Holland (1975) and Fienberg (1977). Let us consider the observed counts x_a from an s dimensional multinomial distribution into a number of mutually exclusive classes (a = 1, 2, ...t) as realisations of random variables X_a with expectations m_a, where $G_a = logm_a$ can be written as a known linear function of a set of unknown parameters $\Sigma K_{ab} \lambda_b$ in a matrix terms denoted by $G = K\lambda$. The columns of the matrix K are the design vectors of co-efficients of each parameter in turn. These design vectors define the model being fitted.

For a closed population with s samples the observations $x_a = x_{ijk...}$ could be represented as a 2^s contingency table with one unobservable category -- a structural zero in one cell. Fienberg (1972) proposed that a model be selected for the observable categories from among the standard hierarchy of log-linear models for contingency tables. With the exception of the unobservable category, the structure of the data is exactly that of a factorial experiment, each factor at two levels (presence or absence, counted or not counted in that sample), which is classically described in terms of a mean (λ), main effect (λ_i), two factor interactions (λ_{ij}), describing how each main effect is changed according to whether or not one other factor is present, three factor interactions (λ_{ijk}) and so on.

Different ways of defining these effects by parameters are possible. Fienberg (1972) used two alternative definitions. According to the first definition, in the full model each parameter represents that effect averaged over both levels of the others, that is, main effects and interactions averaged over all levels of the other factors. According to the second definition the main effect of a particular sample contrasts the number of individuals not counted in that sample but counted in every other with those counts in all samples. The most general model for a complete 2^3 table, according to the second definition is:

$G_{111} = \lambda$

$G_{211} = \lambda + \lambda_1$

$G_{121} = \lambda + \lambda_2$

$G_{221} = \lambda + \lambda_1 + \lambda_2 + \lambda_{12}$

$G_{112} = \lambda + \lambda_3$

$G_{212} = \lambda + \lambda_1 + \lambda_3 + \lambda_{13}$

$G_{122} = \lambda + \lambda_1 + \lambda_3 + \lambda_{23}$

$G_{222} = \lambda + \lambda_1 + \lambda_2 + \lambda_3 + \lambda_{12} + \lambda_{13} + \lambda_{23} + \lambda_{123}$

This has the advantage that with x_{222} unobservable, the model for G_{222} is not defined, so that there are 7 observations modelled naturally by 7 parameters, since λ_{123} does not appear.

If we select a model with no interaction between second and third samples, we get $\lambda_{23} = 0$, and this single constraint is reflected in the relation

$G_{122} + G_{111} = G_{121} + G_{112}$ or $m_{121} / m_{111} = m_{122} / m_{112}$

rather than the more usual relation between the marginal totals $m_{+21}/m_{+11} = m_{+22}/m_{+12}$. In this the subscript '+' denotes summation over possible values of the subscript, for example, $m_{+21} = m_{121} + m_{221}$, the expected number of individuals count in the third sample, but not in the second. The marginal relation is less easy to handle because the fact that x_{222} is unobservable implies that all of x_{2+2}, x_{22+}, x_{+22}, x_{++2}, x_{+2+}, and x_{2++} are unobservable.

5.4 Triple-System Estimation

The incomplete 2^3 table of counts for a triple-system census model is illustrated in Table 5.2. This census model involves eight basic counting statuses corresponding to the eight possible combinations of counts, each of which has two possible outcomes. Thus we have

x_{111} = events reported in all three sources
x_{121} = events reported in sources, first and third
x_{211} = events reported in sources, second and third
x_{221} = events reported in third source only
x_{112} = events reported in sources, first and second
x_{122} = events reported in first source only
x_{212} = events reported in second source only
x_{222} = events not reported in any sources

Table 5.2. Triple-System Data

		Third Sample			
		1		**2**	
		Second Sample		**Second Sample**	
		1	2	1	2
First Sample					
1		x_{111}	x_{121}	x_{211}	x_{221}
2		x_{112}	x_{122}	x_{212}	x_{222}

In terms of this three source model, the statistics x_{11}, x_{12}, x_{21}, and x_{22} of the two sources model become

$x_{11} = x_{11+} = x_{111} + x_{112}$;
$x_{12} = x_{12+} = x_{121} + x_{122}$;
$x_{21} = x_{21+} = x_{211} + x_{212}$; and
$x_{22} = x_{22+} = x_{221} + x_{222}$;

From this three-source matched system we can estimate x_{222} as well as $\check{N}$ in number of ways. For example by averaging the three dual system estimates,

$$\check{N} = 1/3[(x_{1+1} + x_{1+2})(x_{+11} + x_{+12})/ (x_{111} + x_{112}) +$$
$$(x_{1+1} + x_{1+2})(x_{1+1} + x_{2+1})/ (x_{111} + x_{121}) +$$
$$(x_{+11} + x_{+12})(x_{1+1} + x_{2+1})/ (x_{111} + x_{211})] \tag{5.1}$$

or, to produce less bias than the separate ratio estimator, the combined ratio estimator below could be used (Mark et al 1974):

$$\check{N} = 1/3[(x_{1+1} + x_{1+2})(x_{+11} + x_{+12})/ (x_{111} + x_{112})]$$
$$[(x_{1+1} + x_{1+2})(x_{1+1} + x_{2+1})/ (x_{111} + x_{121})]$$
$$[(x_{+11} + x_{+12})(x_{1+1} + x_{2+1})/ (x_{111} + x_{211})] \tag{5.2}$$

We can also divide this triple system count table into one complete 2x2 subtable and one incomplete 2x2 subtable. By assuming that the cross-product ratio k is the same in both subtables, then the cross-product ratio for the incomplete subtable can be estimated from the complete one, as $\check{k} = (x_{111} + x_{221})/(x_{121} + x_{211})$. Applying the DSE to the incomplete subtable, we obtain

$$x_{222} = \check{k} [(x_{122} + x_{212})/(x_{112}) = (x_{111}x_{122}x_{212}x_{221})/(x_{211}x_{121}x_{112}) \tag{5.3}$$

This is equivalent to assuming that $\rho = 1$, where

$$\rho \equiv (p_{111}p_{122}p_{212}p_{221})/(p_{222}p_{211}p_{121}p_{112}) \tag{5.4}$$

(the p's are the cell probabilities); that is, it is equivalent to assuming that there is no second order interaction in Table 5.2. The no second order interaction assumption for the 2^3 table is in one sense analogous to the assumption of independence for the 2x2 table but one layer deeper. All pairs of sources can exhibit dependence, but the amount of dependence in each pair is assumed to be unaffected by conditioning on the third source. Also because the cell x_{222} is missing, the assumption $\rho=1$ is untestable in isolation -- just as k=1 is untestable for the 2x2 table (Darroch et al 1993).

Darroch (1958) first explored the use of three or more lists with a multinomial probabilistic model for the independent lists. His model is

$$pr(n_{\omega}) = [N!/(N - n)!][(1 - \Sigma P_{\omega})^{N-n} \prod_{\omega} P_{\omega}^{n_{\omega}}]/n_{\omega}!] \tag{5.5}$$

In this model it is assumed that

* N is the total number of individuals in the population, which is also the unknown parameter of the model
* n is the total number of different individuals observed in the complete experiment
* ᵚ is the non-empty subset of the integers from l to s where s is the total number of independent lists.
* $P_ᵚ$ is the probability that an individual is caught in the samples corresponding to ᵚ.
* $(1 - \Sigma P_ᵚ)$ is the probability of the null subset
* $(N - n)$ is the number of individuals not observed, where $n = \Sigma n_ᵚ$

Sanathanan (1972) has shown that under suitable regularity conditions both conditional and unconditional maximum likelihood estimators of the parameters of the above equation are consistent estimators and they have the same asymptotic multivariate normal distribution. Following that, and using conditional maximum likelihood estimators, Fienberg (1972) extended Darroch's model into log-linear models to get various estimates of N, the total population size, from the multidimensional incomplete contingency tables.

Using information from three different sources, census (C-source), PES (P-source), and administrative lists (A-source) Zaslavsky and Wolfgang (1990) presented various triple-system estimates of the number of uncounted people, based on log-linear or log-linear like models. They estimated the uncounted people from the full three system table as well as from various marginal subtables of the three system table. These models are,

1 **C-D estimate without A-source:** In this model A-source information was not used, that is, it is an ordinary C-D estimator based on C-- and P-sources. The C-D estimate is $[x_{22+} = x_{12+}x_{21+} / x_{11+}]$. x_{222} is obtained by subtracting x_{221} from the dual system estimate x_{22+}.

2. **C-D estimates with P+A:** In this model information from the P-source is combined with the A-source to make a single second source. The C-D estimate is
$$x_{222} = x_{122}(x_{21+} + x_{221}) / (x_{11+} + x_{121})]$$

3. **C-D estimates with k_2:** In this model it is assumed that the degree of dependence between the C-- and P-sources is similar in the overall population to that in the subpopulation captured by the administrative lists. Hence, CxP cross product ratio k_2 is estimated from the cells with a=1, $k_2 = (x_{221}x_{111}) / (x_{211}x_{121})$.
The C-D estimate is recalculated by $[x_{22+} = x_{21+} + x_{12+}) / (x_{11+})$.

4. **Estimate from Ratio r_1:** In this model it is assumed that the probability of coverage in the A-source of persons omitted from the C-- and P-sources is the same as the average probability of coverage for those included in at least one of those sources. Hence, an estimate of the odds ratio 'r_1', for coverage by the A-source, based on all cells enumerated in the E-- or P-sources, is obtained by $r_1 = (x_{111} + x_{121} + x_{211})/(x_{112} + x_{122} + x_{212})$. The estimate of the uncounted cell x_{222} is obtained from this odds ratio by applying x_{221}, which is $x_{222} = x_{221} / r_1$.

5. Estimate from Ratio r_2 : In this model it is assumed that the probability of coverage in the A-source of persons omitted from the C- and P-sources is the same as the average probability of coverage for those included in either the C-source or the P-source but not both, but that the persons enumerated in both the E- and P-sources are not necessarily comparable in this respect. In other words, the persons captured by neither source are more like those captured by one than those captured by both. Hence, an estimate of the odds ratio 'r_2', for coverage by the A-source, based on the cells enumerated in E- or P-source, but not both, is obtained by $r_2 = (x_{121} + x_{211})/(x_{122} + x_{212})$. The estimate of the uncounted cell x_{222} is obtained from this odds ratio by applying x_{221}, which is $x_{222} = x_{221} / r_2$.

6. Estimate by assuming $k_3 = 1$: In this model it is assume that the 3-way cross-product ratio k_3 is equal to one. This means that there is no second order interaction in the model, so $x_{222} = (x_{122}x_{212}x_{221}x_{111})/(x_{112}x_{121})x_{1211})$.

5.5 Models for Varying Catchability

Darroch et al. (1993) present some models that allow for varying catchability of individuals as well as varying levels of penetration into the target population of each sample or list. In the smallest possible subdivision of the study population, they assume that the lists are independent within individual, but different individuals may in general have different probabilities of capture. Following the Zaslavsky and Wolfgang (1990, 1993) post strata technique, they then combine these single-individual strata to construct more realistic strata. These resulting combined counts show a positive correlation due to heterogeneity as described by Kadane et al (1992). The resulting log-linear model for the combined strata contains parameters that represent both list effects and the different 'catch efforts' of the sample producing the list.

We now suppose that the $\mathbf{J} = (j_1, j_2, j_3)$ represents the capture pattern: $j_{hi} = 1$ if the individual h is on list i and 0 otherwise. We assume a fixed closed population of size N, where each individual h, for h = 1,2,3,...N, has his or her own fixed catchability parameters. We also consider the hypothetical repetition of the entire triple-system estimation experiment under independent identical conditions.

5.5.1 Independence Model

We are interested in estimating census undercount and according to the condition of the model assume that the three lists are independent. We assume independence across individuals and let us suppose that each individual h has probability $p_h(\mathbf{J})$ of capture pattern $\mathbf{J}$, and actually experiences capture pattern $\mathbf{J}_h = (j_{h1}, j_{h2}, j_{h3})$.

$$p_h(\mathbf{J}) = Л_{1j1}(h)Л_{2j2}(h)Л_{3j3} = \prod Л_{i1}(h)Л_{i0}(h)^{1-h} \tag{5.6}$$

where $\Pi_{i1}(h) = 1 - \Pi_{i0}(h)$ is the probability that individual h is on list i.

The assumption of homogeneous catchability means that the probability of being on each list is independent of h: $\Pi_{i1}(h) \equiv \Pi_{i1}$. Letting $\beta_i = \log \Pi_{i1} / \Pi_{i0}$, the probability $p(\mathbf{J})$ of observing the response pattern $\mathbf{J} = (j_1, j_2, j_3)$ is

$$\log p(\mathbf{J}) = \alpha + j_1\beta_1 + j_2\beta_2 + j_3\beta_3 \qquad (5.7)$$

which is the model of independence for the table x_j.

Now instead of the homogeneous catchability assumption, let us suppose that the individuals have heterogeneous catchability, so that the $\Pi_{i1}(h)$ are allowed to depend on h. Continuing to allow for heterogeneity in the catchability of individuals, we assume that the pattern of heterogeneity is the same for all three samples. More precisely, we assume that for any two individuals h and h', the odds ratio

$$\Pi_{i1}(h)\Pi_{i0}(h') / \Pi_{i1}(h')\Pi_{i0}(h) \qquad (5.8)$$

is constant with respect to i. This assumption is equivalent to the additive-logit model

$$\log[\Pi_{i1}(h) / \Pi_{i0}(h)] = t_n + \beta_i \qquad (5.9)$$

so that capture probabilities are characterized by the logistic function

$$\Pi_{i1}(t) = e^{t+\beta i} / 1 + e^{t+\beta i}$$

5.5.2 No Second Order Interaction Model

Sanathanan (1972a, 1972b, 1973) considers conditional estimation of the population size N for the DSE situation and shows that $\check{N}_u \leq \check{N}_c$ (u = unconditional, c = conditional) and that, under suitable regularity conditions, both conditional and unconditional approaches provide consistent estimators and have the same asymtotical multinomial normal distribution. Following Sanathanan (1972a) and Fienberg (1972), Darroch et al. (1993) analyze the incomplete 2^3 table conditionally. Thus instead of estimating parameters directly from the likelihood (5.15), they work with the likelihood based on the conditional probability of the observable frequencies, given $n = x_{221} + x_{212} + x_{211} + x_{122} + x_{121} + x_{112} + x_{111}$; that is,

$$n! \, \Pi_{(j1, j2, j3) \neq (2,2,2)} \, [\{p_{j1, j2, j3} / (1 - p_{222})\}^{x j1 j2 j3}] / x_{j1j2j3}! \qquad (5.10)$$

When n is given by using (5.10) one can estimate model parameters. Once the parameters have been estimated one must be able to write the cell probability p_{222} in terms of these parameters in order to generate an estimate x_{222} for the unobserved cell count.

The quasi-symmetry model (5.17) for j = 3 lists is equivalent to the constraints

$$p(222)p(122)= p(121)p(212) = p(112)p(221) \tag{5.11}$$

and does not relate the probability p(222) to the other seven probabilities. Thus an additional assumption, such as no second-order interaction, is needed. Under the Rasch/quasi-symmetry model (5.17), the no second-order interaction model (5.4) becomes

$$\rho = e^{\gamma(3)}e^{3\gamma(1)}/e^{\gamma(0)}e^{3\gamma(2)} \tag{5.12}$$

where $\gamma(k)$ is defined as in (5.18). Hence the model (5.17) reduces to

$$\log p(\mathbf{J}) = \alpha + j_1\beta_1 + j_2\beta_2 + j_3\beta_3 + \gamma(j_+)^2 \tag{5.13}$$

where α, β's and γ are all linear coefficients, with $\gamma > 0$.

5.5.3 Quasi-symmetry Model

For the larger N^2 table $w_{h,j}$ the cell probability is equal to the quantity $p_h(\mathbf{J}) = \Pi_{1j1}(h)\Pi_{2j2}(h)$ $\Pi_{3j3}(h)$ when $\mathbf{J} = (j_1, j_2, j_3)$ represents the capture pattern of an individual $j_i = 1$ if the individual is on list i and 0 otherwise. It is easy to see that the cell probabilities for the marginal 2^3 table x_j must be

$$p_h(\mathbf{J}) = = p_{j1j2j3} = 1/N\Sigma_{h=1}^{N} \Pi_{1j1}(h)\Pi_{2j2}(h)\Pi_{3j3}(h) \tag{5.14}$$

In many situations especially when there are no coverage errors with respect to the scope of the area and/or time period in which the events are recorded and when there are no misclassification errors with respect to determining whether a particular event has been recorded by all three information sources or two or only one of them, the counts x_J will approximately follow the multinomial distribution (El-Khorazaty et al 1977)

$$N! \, \Pi(P_{j1j2j3}^{\,xj1j2j3})/(x_{j1j2j3})! \tag{5.15}$$

We may rewrite the cell probabilities in (5.14) by using (5.6) and (5.9) as

$$p_h(\mathbf{J}) = 1/N\Sigma_{h=1}^{N} \Pi_{i=1}^{3} \, [\Pi_{i1}(h) / \Pi_{i0}(h)] \, \Pi_{i0}(h)$$

$$= 1/N\Sigma_{h=1}^{N} \prod_{i=1}^{3} e^{ji(th+\beta i)} \Pi_{i0}(h)$$

$$= \exp[j_1\beta_1 + j_2\beta_2 + j_3\beta_3] \, 1/N\Sigma_{h=1}^{N} [e^{th}]^{j+} \, ph(\mathbf{0}) \tag{5.16}$$

taking logarithms we obtain a log-linear model for the 2^3 table x_j of the form

$$\log p(\mathbf{J}) = \alpha + j_1\beta_1 + j_2\beta_2 + j_3\beta_3 + \gamma(j_+) \tag{5.17}$$

where $J_+ = j_1 + j_2 + j_3$. It follows from Golland (1990a) that

$$\gamma(k) = \log E[e^{kt}|J = 0] \tag{5.18}$$

where T follows the posterior distribution of the catchability effects t conditional on not being caught in any sample $\mathbf{J} = \mathbf{0}$.

5.5.4 Partial Quasi-symmetry Model

In equation (5.13) we have five parameters, α, β_1, β_2 β_3 and γ and in the quasi- symmetry and no second-order interaction models we have seven observed cell probabilities, $p_{111},\ldots\ldots p_{212}$, leaving 2 degrees of freedom for assessing model fit. The deviances of quasi-symmetry can be seen from the three frequency products corresponding to the probability products in (5.11). Darroch et al. (1993) observed consistently large differences between $x_{112}x_{221}$ and the other two products, $x_{211}x_{122}$ and $x_{121}x_{212}$. They comment, for several of the tables, the products $x_{211}x_{122}$ and $x_{121}x_{212}$ are fairly close together. Thus it seems reasonable to assume that

$$p(211)p(122) = p(121)p(212) \tag{5.19}$$

Property (5.19) may be interpreted in terms of the individual capture logits (5.9): Equation (5.19) arises by assuming that

$$\log[\Pi_{i1}(h) / \Pi_{i2}(h)] = t_n + \beta_i, \; i = 1,2 \tag{5.20}$$

$$= s_h + \beta_3 \; i = 3 \tag{5.21}$$

That is, (5.19) arises from the assumption (5.19) that the pattern of heterogeneity is the same for the first two samples; census and CVS samples only (and different in the third sample). Hence, following the arguments of quasi-symmetry model, we get a ``partial quasi-symmetry" model for the table x_j, replacing (5.17):

$$\log p(\mathbf{J}) = \alpha + j_1\beta_1 + j_2\beta_2 + j_3\beta_3 + \gamma(j_1 + j_2 + j_3) \tag{5.22}$$

where $\gamma(k_1, k_2) = \log E[U^{k_1} V^{k_2} | \mathbf{J} = 0]$ for positive random variables $U = e^T$ and $V = e^S$; compare (5.18).

5.6 Result

We consider four models:

1. the model of complete independence among the three lists
2. no-second-order-interaction model
3. submodel assuming full quasi-symmetry
4. the submodel assuming partial quasi-symmetry

Our main interest is to observe which of the above models fit the data well. In all the four strata the complete independence model fits poorly. In stratum 1 except row 1 (Table 5.3) the estimated values by the three models were similar. In row 1 the estimated values are higher than the other four rows of the stratum and also the full quasi symmetry model gives higher values than no-second-order interaction and partial quasi symmetry models. In stratum 2 partial quasi symmetry model provides higher values than no-second-order interaction model while the full quasi symmetry model fits best. In this stratum the estimated values of rows 1 and 5 are much smaller than the other three rows. In stratum 3 the no-second-order interaction model provides comparable improvements in fit over the partial quasi symmetry model, except in row 3 while again the full quasi symmetry model fits best. In stratum 4 the results are a little bit different than the other three strata. Here, values from quasi and partial quasi symmetry models are more similar than the no-second-order interaction model. In this stratum the partial quasi symmetry model provides a much better fit than does the no-second-order interaction model; and the quasi symmetry model provides comparable improvements in fit over the partial quasi symmetry model. In rows 1, 4 and 5, however, the no- second-order interaction model does not fit well while in row 1 the partial quasi symmetry model fit poorly.

5.7 Conclusions

In this Chapter we have discussed some of the models presented by Zaslavsky and Wolfgang (1990) and Darroch et al. (1993). We considered four strata and three different sources of information, census, CVS and administrative lists, to see how the estimate of x_{222} is affected by the additional information from each of the above sources of information. In stratum 1, we considered that the information from the three sources was more or less the same. Results (Table 5.3) shows that the estimated values of row 1 are much more higher than the other four rows. One of the reasons for the higher estimated values in row 1 compared to the other estimated values of the stratum may be that, though we have more or less same information from the three sources, in this particular case, CVS gives a little bit more information than the other two sources. The estimates from the no-second-order-interaction and the partial quasi symmetry models are similar while estimates from the quasi symmetry model are higher than all the other three models. This means that, if we do not have any additional information from any of the sources of information the no-second-order-interaction and the partial quasi symmetry have no advantage over one another. However, the quasi symmetry model may still have some advantages.

In stratum 2 we considered that the census counts and CVS estimates give similar information while the administrative list gives different and higher number of individuals than both the census and the CVS. From the results we observed that all the three models give much higher values than the complete independence model. This clearly reflects the assumption that the individuals have varying catchability. Comparison among no-second-order-interaction, partial quasi symmetry and quasi symmetry models shows advantages of the partial quasi symmetry model over the no-second-order-interaction model and the quasi symmetry model over the partial quasi symmetry model. One can also observe from the results of the quasi and the partial quasi symmetry models that in row 4 and row 5 values are almost equal, which is exceptional compare to the whole results of the Table 5.3. One reason may be that when all the three sources of information provide additional information (Table 5.1) the partial quasi symmetry model fits the data best.

Finally, when we distributed the matching results from the three sources of information into the 7 cells of a 2x2x2 table arbitrarily, we were concerned only with the additional information from the third source. During fitting the model we realised that each of the cell frequencies of the table may have some affect on the estimates. For example, one can observe from the estimated values by quasi and partial quasi symmetry models of row 1 and row 5 of stratum 2, that the value of row 5 is greater than row 1 by 100 (approximately). The data from which these values were estimated (Table 5.1) shows that the frequencies in all the cells for these two estimates were almost equal except in two. Hence, any straightforward conclusion from these hypothetical data is difficult.

Table 5.3. Log-linear Model Estimates for x_{222}

Stratum 1	Complete Indep.	No Second Order Inter.	Full Quasi Symmetry	Partial Quasi Symmetry
1	0.617	742.982	984.139	731.373
2	0.579	211.200	232.481	210.182
3	0.418	229.091	243.300	229.365
4	0.566	411.136	462.501	402.000
5	0.396	117.385	136.250	117.720
Stratum 2				
1	1.555	294.171	554.024	428.060
2	2.334	1135.200	1760.123	1395.997
3	2.860	1166.080	1350.750	1322.700
4	4.110	1045.870	1134.010	1139.340
5	2.580	372.690	527.690	528.180
Stratum 3				
1	2.320	1477.770	3497.540	1476.010
2	2.720	581.770	1119.790	555.840
3	3.510	1164.890	2253.680	1198.310
4	4.570	1546.150	2232.460	1540.780
5	2.950	545.250	1051.370	498.510
Stratum 4				
1	1.900	387.025	1414.357	653.452
2	3.820	2547.727	3687.261	3370.637
3	3.560	1909.500	3718.982	3504.032
4	3.000	988.060	1821.520	1488.760
5	4.090	991.420	1694.370	1384.720

Chapter 6

Regression Models to Estimate the Local Population

6.1 Introduction

Population censuses try but typically fail to count everyone. Any census will underestimate individuals and households, and may also overestimate. Even when household units are identified, occupants may be missed when they are not reported to the census. Still other people are missed because they have no usual place of residence.

The demand of adjusting census counts and the theory and methods that might be used for the purpose have been matters of hot dispute for at least last forty years. Starting with the basic demographic analysis techniques proposed by Coale(1955, the approaches used have progressively synthesised more statistical approaches which bring in a stochastic element. The need for more accurate census counts arose because of the increased uses of census data in allocating Central Government funds, in public and private planning, and in determining the eligibility of a locality for funding or government programs (Edmonston et al, 1995). The undercount became a particular target of concern, because estimates provided by the Census Office show that it is more concentrated among men than women, and among the young rather than the old.

In this chapter we are concerned with the underenumeration in the U.K. 1991 Population Census. Comparison with demographic estimates shows that 1.2 million people were missed from the 1991 census counts (Census Newsletter No. 24, OPCS, 1992). Certain groups of people, characterised either by their demographic, spatial, socio-economic or household features are likely to have greater or smaller probabilities of being underenumerated than the average. The main aim of adjusting for underenumeration is to decide whether to allocate the underenumeration disproportionately among certain subgroups of the population and, if so, how should this be done (Diamond, 1993).

In the following we present a method of adjusting census population counts for small areas. We begin with a description of the regression models. A discussion of estimating undercount rates which are used as dependent variables in the regression equations is then described. Succeeding sections describe the adjustment results and summarize our findings.

6.2 Regression Models

Regression can be applied using estimates from post enumeration surveys and census counts for local areas. For each area i where sample data are available, we would have y_i, the sample estimate, of the local population as the dependent variable in the regression. For independent variables, symptomatic indicators of each area can be used. Independent variables might be the percentage of male/female persons present in the household on the census night, characteristics of migration, characteristics of age groups or various economic indicators. It would not be necessary to use variables from the census itself except for the local counts stratified by age, race and sex. To reduce both the variability and skewness of the distribution all the variables were written in percentages (Ericksen, 1974).

The main contribution of the regression method to the current measurement of the undercount rate or population total for the local areas is that regression estimates could be calculated both for those areas where sample data were available and for those remaining areas were sample data were not available. Ericksen (1974), applied this technique to compute a regression equation estimating population growth from sample data in several hundred primary sampling units (PSU's) included in the Current Population Survey (CPS) in the U.S. and then used the equation to estimate population growth in 2586 counties in 42 states. He claimed that in the absence of correlation between the sampling variation and the true values, the estimates of the regression coefficients are unbiased, although the correlations have decreased and the mean square error of the regression estimates increased. To increase the quality of the regression estimates he suggests stratifying the population and computing separate equations for different categories of areas such as central cities, suburbs, and metropolitan areas.

To compute local estimates of undercount we will fit regression models as suggested by Ericksen and Kadane (1985). To specify the models, if we assume that U_i denotes the undercount estimates for Local Authority i, a regression can be run on a variety of variables, such as the percentage of the population over 60, X_{1i}, the percentage of males, X_{2i}, and percentage of census data imputed, X_{3i}, with the equation

$$U_i = a + b_1 X_{1i} + b_2 X_{2i} + b_3 X_{3i} + e_i \qquad (6.1)$$

where

a = the intercept of the regression.

b_1, b_2, and b_3 = regression coefficients for the age, sex and imputation variable respectively.

e_i = random error of Local Authority i.

To go to lower levels of Local Authority, such as Ward, j, within Local authority Areas, the allocation would be implemented as

$$\hat{U}^*_{ij} = a + b_1X_{1ij} + b_2X_{2ij} + b_3X_{3ij} \qquad (6.2)$$

6.2.1 Assumptions of the Models

1. the estimated undercount for Local Authority Area i is equal to the true undercount for that area.
2. the undercount estimates are unbiased.
3. estimates are unrelated from area to area.
4. true census errors are linearly related to a set of explanatory variables.
5. the error terms are normally distributed
6. the expected values of the error terms in the regression equations are zero.
7. error terms have equal variances.

This regression sample data technique has been tested in a variety of empirical situations. Gonzalez and Hoza (1978), estimate unemployment of the U.S. by using Current Population Survey estimates as a dependent variable and independent variables obtained from administrative sources, census data and synthetic estimates. Stell and Poulton (1988), have conducted a regression analysis using the 157 local government authorities (LGAs) in the major cities of Sydney, Melbourne, Adelaide and Perth. The dependent variable was the LGA underenumeration rate estimated directly from the post-enumeration survey. Independent variables were collected from the census enumeration. Both unweighted and weighted linear regression models were fitted, among which weighted regression has the intuitively appealing property of reducing the influence of LGAs with small sample sizes. The estimated LGA underenumeration rates vary from -0.013 to 0.104. Ericksen and Kadane (1985), discussed the effect of the three main assumptions of the dual-system methods when they fail in estimating underenumeration of the U.S. population and suggested different ways of overcoming the problems and estimation procedures of the more accurate underenumeration rate of the U.S. census counts which can be used as a dependent variable in the regression models. Isaki et al (1988), give synthetic estimates in combination with a regression model to estimate census undercount in small areas. In the U.K. the CVS has a completely different methodology for estimating the undercount rate. Based on this different situation and using demographic estimates OPCS estimate the local area undercount rate in a different way. This is discussed in the following section.

6.3 Estimating the Undercount Rate

The 1991 demographic estimates were based on the 1981 Census counts (adjusted for coverage error), birth and death records, and immigration and emigration data. For 1991 the demographic estimates are calculated very simply using the classic balancing equation in which 1981 census counts were used as baseline estimates (adjusted for coverage error), with recorded births and recorded deaths data used to estimate the natural growth of the population in the last ten years and immigration and emigration data used for estimating the net effect of migration at the national level (2.2, Chapter 2). When all the sources of information are reliable or accurate, demographic estimates are not only the simplest but are also very reliable. However complications can occur if any of the data sets fails to provide accurate or up-dated information. It is, therefore, important to check the possible sources of error in the demographic estimates. In the case of the 1991 demographic estimates, the registration of births and deaths is believed to be virtually complete (Heady et al. 1994). It is unlikely that the registration of total births at the national level is too high or the registered number of total deaths is too few. From the International Passenger Survey (IPS) and other ancillary data it was felt that the immigration figure was more likely to understate than overstate immigration (Heady et al, 1994). The 1981 methodology of the Post Enumeration Survey (PES) and the methodology of the 1991 Census Validation Survey (CVS) were more or less the same. If the PES overestimated the undercount rate, then it is also possible and believable that the 1991 CVS also overestimated the undercount rate. But that is not the case, so the base estimate which is used in the demographic estimates does not overestimate the demographic estimates significantly.

Comparison between the demographic estimates and the 1991 census counts shows that the Census total was similar to the demographic estimates for the age groups 45-79. The greatest differences observed even after correction by CVS were for the age groups 20-34. It was also observed that underenumeration was much more among young men than among young women. Beside this, it is expected that the extent of underenumeration is higher in Inner London, the Main Metropolitan cities and Non-metropolitan cities than in other parts of the country (Diamond, 1993). This suggested that the ratio of enumerated males to females in the age group 20-34 would be lower than one would expect in areas with particularly high levels of under enumeration. As the underenumeration in different areas is not the same and the enumerated sex ratios of males to females in the age group 20-34 are lower than expected we believed it was appropriate to use the difference between the Census sex ratio for each type of areas and the corresponding expected sex ratio as an indicator of differential underenumeration. Based on the OPCS methodology (Heady et al, 1994) it was therefore decided to estimate three sets of underenumeration rates on the basis of the sex ratios of the three age groups, 20-24, 25-29, and 30-34, for the 403 Local Authorities (LA) of the U.K..

To estimate the undercount rate local authorities were divided into six area type categories -- namely Inner London, Outer London, main metropolitan cities, other metropolitan areas, non-metropolitan cities and other metropolitan areas. By using sex ratios we calculate the expected sex ratio of each LA by assuming that the relation between area type sex ratio to the national sex ratio was the same in 1991 as it has been in the past. More specifically we assumed that the relationship in 1991 was the average of those in 1971 and 1981, and calculate the expected sex ratio which is

$$S_a = S_{91n}[(S_{71a}/S_{71n}) + (S_{81a}/S_{81n})]1/2 \qquad (6.3)$$

where
S_a = expected sex ratio for local area
S_{91n} = 1991 national sex ratio
S_{71n} = 1971 national sex ratio
S_{81n} = 1981 national sex ratio
S_{71a} = 1971 sex ratio for local area
S_{81a} = 1981 sex ratio for local area

These expected sex ratios are then compared with the sex ratios of each area. To balance the equation of the two sex ratios, that is, the local area sex ratio and expected sex ratio we used the factors as follows

$$S_a = M(1 + mk)/F(1 + fk) \qquad (6.4)$$

$$k = (M-S_a F)/ (S_a Ff-Mm) \qquad (6.5)$$

where

S_a = expected sex ratio for local area
M = Male total so far for local area
F = Female total so far for local area
m = male national under-enumeration
f = female national under-enumeration
k = undercoverage indicator

The k of the equation (6.4) is the key factor, which balances the equation and hence is described as an indicator of the undercoverage rate of the local area with that of national average. This indicator of undercoverage is applied to both the male and female undercoverage rate with the

75

assumption that in any given local area the male and female underenumeration rates differ from the national average in the same direction and by the same proportionate amount. We estimated Ks from equation (6.5) for the age groups 20-24, 25-29, and 30-34 and used these as the dependent variables in the regression equation.

6.4 Procedure for fitting the Regression Models

For estimating the Local Authority Population Total our procedure for fitting the regression models is as follows:

A. The underenumeration rate estimated directly from the CVS sample for the 403 Local Authority Areas by equation (6.5) is used as the dependent variable in the regression models.

B. We collected symptomatic information from the SASPAC (Small Areas Statistics Package), for each local authority. These Local Authority Areas are the minimum level for which we fitted the regression models.

C. We collected symptomatic information as independent variables for the regression model in the form of percentages as it is convenient to use for our case.

D. We fitted the regression models using SPSSPC+ to estimate the parameters for the Local Authority Areas, as the underenumeration rates (dependent variable) are available only for Local Authority. Values of the symptomatic indicators are then substituted into this equation to obtain estimates for each Electoral Ward's population.

E. To select the variables with largest positive or negative correlation with the dependent variable we used a forward and stepwise selection approach. Both approaches selected the same variables for the model.

F. The potential independent variables which were included in the regression analysis were collected from the Australian experience and are given in Appendix 3.

6.5 Selecting Explanatory Variables

The level of census undercount varies from place to place (Ericksen, 1980). Many statisticians believe that these variations highly depend on the characteristics of the local population and hence increase the pressure to consider local population characteristics while estimating the local population. Ericksen (1973) found that changes in symptomatic variables such as births, deaths and school enrollment are useful indicators of changes in the size of local population. In 1985, he used the local population characteristics in the regression equation as explanatory variables to estimate the undercount rate of local areas such as city or state in the U.S.. Freedman and Navidi (1986) also followed the same philosophy and used eight explanatory variables to built their regression model. We also believe that undercount rate of a particular age group depends on the characteristics of that age group as well as on the percentage of population of that group. Keeping this in mind in our models we began the analysis with thirty five explanatory variables from SASPAC and built three regression models using the subset of seven, eight and ten variables respectively that best fit the estimated undercount rate by the ordinary least squares. Some of these explanatory variables are:

1. X_{01} : Percentage of male,

2. X_{03} : Percentage of black and other residents,

3. X_{04} : Percentage of persons born in NewCommonwealth and Africa

4. X_{07} : Percentage of male persons in the age group 20-24,

5. X_{08} : Percentage of residents with limiting long-term illness in the age group 16-44,

6. X_{09} : Percentage of economically active residents with limiting long- term illness in the age group 16-44,

7. X_{10} : Percentage of male migrants between district but within county in the age group 25-29,

8. X_{12} : Percentage of International migrants, etc..

6.6 Estimating the Regression Equation

Using all the LA areas of England and Wales a model was selected by using a forward selection approach. The model selected only six independent variables from the 35 variables.

1. X_3 = Percentage of black and other residents,
2. X_{12} = Percentage of migrants from outside UK,
3. X_{21} = Percentage of persons in the age group 20-24,
4. X_{30} = Percentage of NewCommonwealth households with residents,
5. X_{32} = Percentage of households used as a second residence and
6. X_{33} = Percentage of persons in the age group 16-19.

Among the six selected variables listed above, the variables X_3 and X_{30} were highly correlated (r = 0.973) with each other. In general, any large intercorrelations between the independent variables indicate the presence of multicollinearity (Ericksen, 1973). One of the simple way of handling multicollinearity if found, is to delete the offending variable/variables from the analysis. The information may not be lost by deleting the offending variable since it is combination of other variables (Berk, 1977). Frane (1977) suggest use of stepwise, or hierarchical entry of variables into the analysis so that only one or a few of the variables that are multicollinear are used. We therefore, removed the variable X_{30} from the analysis and reselected the variables for the regression model. The model selected the following seven variables.

1. X_4 = Percentage of persons born in NewCommonwealth and Africa.
2. X_{12} = Percentage of migrants from outside UK.
3. X_{21} = Percentage of persons in the age group 20-24.
4. X_{27} = Percentage of households with male lone pensioners in the age group 65- 74.
5. X_{32} = Percentage of households used as a second residence.
6. X_{33} = Percentage of persons in the age group 16-19 and
7. X_{34} = Percentage of persons in the age group 30-44.

The regression estimates of the local population in which the undercount rate estimated from the sex ratio for the age group 20-24 (using equation (6.5)) was used as the dependent variable and the above seven variables as independent variables can be represented by the following equation:

$$\hat{Y}_1 = 2.10 - .067X_4 + .154X_{12} + .033X_{21} + .28X_{27} + .097X_{32} + .785X_{33} - .137X_{34} \qquad (6.6)$$

where

$\hat{Y}_1$ is the estimated undercount rate for the age group 20-24. The R Square and the adjusted R Square of the model are 0.32903 and 0.31886 which indicate that about 33 per cent variation is explained, and the fit of the model was reasonably satisfactory. The standard errors of the estimates are shown in Table 6.1.

Table 6.1: Estimates and Their Standard Errors (Model 1)

Variables	Estimate	Standard Error	Sig T
X_4	-0.0670	0.0205	0.0012
X_{12}	0.1540	0.0193	0.0000
X_{21}	0.0327	0.1230	0.7902
X_{27}	0.2860	0.0824	0.0006
X_{32}	0.0968	0.0568	0.0863
X_{33}	0.7847	0.2568	0.0024
X_{34}	-0.1371	0.0495	0.0059
CONSTANT	2.0992	1.5484	0.1760

The correlation matrix in Table 6.2 show that the correlations between any two variables is neither too high or too low. Highest correlation was observed between the percentage of persons born in NewCommonwealth and Africa and the percentage of International migrants (r = 0.53).

Table 6.2: Correlation Matrix of Model 1

	X_4	X_{12}	X_{21}	X_{27}	X_{31}	X_{32}	X_{33}	Y
X_4	–	.531	.455	.345	-.108	.395	.030	.220
X_{12}		–	.236	.144	.075	.242	.039	.396
X_{21}				.577	-.161	.646	-.022	.399
X_{27}					-.055	.475	-.113	.348
X_{32}						-.021	-.485	.123
X_{33}							-.167	.396
X_{34}								-.145

We fitted a second regression model where dependent variables were the estimated undercount rate for the age groups 25-29. These undercount rate were estimated from the sex ratio by using the equation (6.5). The model selected eight variables which we used to estimate the local population. Our regression equation is:

$$\hat{Y}_2 = 7.257 - .331X_1 + .317X_7 + .193X_8 + .176X_9 + .047X_{12} + .152X_{17} + .432X_{27} - .079X_{34}$$

(6.7)

Where

$\hat{Y}_2$ is the estimated undercount rate for the age group 25-29.

X_1 is the percentage of the male persons present in the households on the census night.

X_7 is the percentage of the male persons present in the households in the age group 20-24 on the census night.

X_8 is the percentage of residents in the households with limiting long-term illness in the age groups 16-44.

X_9 is the percentage of economically active residents in the households ith limiting long-term illness in the age groups 16-44.

X_{17} is the percentage of dependent persons in the households with residents.

X_{27} is the percentage of households with male lone pensioners in the age group 65-74.

X_{34} is the percentage of persons in the age group 16-19.

The R Square and the adjusted R Square of the model are 0.401 and 0.389 which indicate that about 40 per cent variation is explained, and the fit of the model was reasonably satisfactory. The standard errors of the estimates are shown in Table 6.3.

Table 6.3: Estimates and Their Standard Errors (Model 2)

Variable	Estimate	Standard Error	Sig T
X_1	-0.331	0.052	0.0000
X_7	0.317	0.086	0.0003
X_8	0.192	0.043	0.0000
X_9	0.176	0.056	0.0019
X_{12}	0.047	0.013	0.0002
X_{17}	0.152	0.027	0.0000
X_{27}	0.432	0.048	0.0000
X_{34}	0.079	0.027	0.0031
Constant	7.257	2.102	0.0006

The correlation matrix for regression model 2 in Table 6.4 shows that none of the variables are highly correlated. However, the highest correlation was observed between the variables X_1 and X_{34}, that is, between the variables 'percentage of male persons' and `persons in the age group 30-44'(r = 0.65).

Table 6.4: Correlation Matrix of Model 2

	X_1	X_7	X_8	X_9	X_{12}	X_{17}	X_{27}	X_{34}	Y
X_1		.181	.291	.262	-.283	.194	-.077	.649	-.256
X_7			.603	.390	.148	.189	.392	.111	.266
X_8				.543	.173	.452	.506	.472	.281
X_9					.459	-.354	.094	.561	.102
X_{12}						-.425	.144	.039	.260
X_{17}							.314	.072	.194
X_{27}								.113	.501
X_{34}									-.075

A third model was fitted by using the undercount rate for the age group 30-34 as the dependent variable. This model selected ten independent variables and we used all the variable to estimate the local population total. The regression equation is as follows:

$$\hat{Y}_3 = 30.233 - .647X_1 - .346X_{10} + .139X_{12} - .179X_{13} - .449X_{15} - .071X_{16} + .903X_{27} - .561X_{28} + .132X_{31} + .331X_{34} \tag{6.8}$$

Where

$\hat{Y}_3$ is the estimated undercount rate for the age group 30-34.

X_{10} is the percentage of male migrants (age group 25-29) between districts but within county.

X_{13} is the percentage of males imputed in the wholly absent households.

X_{15} is the percentage of households having three or more cars.

X_{16} is the percentage of households having five rooms.

X_{28} is the percentage of households with female lone pensioners in the age group 60-74.

X_{31} is the percentage of birth (inside UK) of households head of the NewCommonwealth residents.

The R Square and the adjusted R Square of the model are 0.297 and 0.279 which indicate that about 30 per cent variation is explained, and the fit of the model was reasonably satisfactory. The standard errors of the estimates are shown in Table 6.3.

Table 6.5: Estimates and Their Standard Errors (Model 3)

Variables	Estimates	Stand Error	Sig T
X_1	- 0.647	0.144	0.0000
X_{10}	-0.346	0.137	0.0121
X_{12}	0.139	0.034	0.0001
X13	-0.179	0.054	0.0010
X_{15}	-0.449	0.077	0.0000
X_{16}	-0.071	0.029	0.0148
X_{27}	0.903	0.142	0.0000
X_{28}	-0.561	0.106	0.0000
X_{31}	0.132	0.030	0.0000
X_{34}	0.331	0.058	0.0000
Constant	30.233	6.703	0.0000

Some pairs of independent variables used in the regression model 3 have a high correlation (Table 6.6). The highest correlation (r = 0.697) was observed between the variables X_{15} and X_{28}, that is, between `percentage of households having three or more cars' and `female lone parents in the age group 60-74' followed by the variables X_1 and X_{34}, that is, between `percentage of male persons present in the households' and `percentage of persons in the age group 30-44'. The observed correlation between this pair of variables were (r = 0.649).

All three models reflect the strong influence of the three variables, X_{12}, X_{27} and X_{34}, that is percentage of International migrants, percentage of households with male lone pensioners in the age group 65-74 and the percentage of persons in the age group 30-44. Model 1 mainly included two type of variables, first migrant and second percentage of persons of the young age groups and probably reflects the higher underenumeration rate of people away from their place of usual residence at census night.

Model 2 includes mainly percentage of young male persons present in the census night along with percentage of economically active young persons with limiting long-term illness. It seems that the inclusion of two variables, percentage of economically active/not active residents with limiting long term illness and the Percentage of dependent persons in the households improve this model.

Among the three models, Model 3 selected maximum number of independent variables for the regression equation. The inclusion of the variables percentage of males imputed in the wholly absent households, percentage of households having three or more car and percentage of households having five rooms improve the model. However, like the other two Models, this model also probably reflects the higher underenumeration rate of people away from their place of usual residence at census night.

Table 6.6: Correlation Matrix of Model 3

	X_1	X_{10}	X_{12}	X_{13}	X_{15}	X_{16}	X_{27}	X_{28}	X_{31}	X_{34}	Y
X_1		-.006	-.283	.250	-.287	-.270	-.077	-.242	.294	.649	-.122
X_{10}			.239	.114	.054	.025	.097	-.021	-.303	.103	-.095
X_{12}				.168	.097	-.004	.144	.002	-.570	.039	.166
X_{13}					-.030	-.107	.259	.082	-.183	.293	-.043
X_{15}						-.502	-.609	-.697	.274	.292	-.228
X_{16}							.321	.386	-.129	-.297	.023
X_{27}								.635	-.329	-.113	.267
X_{28}									-.293	-.054	.098
X_{31}										.162	.034
X_{34}											.463

6.7 Assessment

To assess the estimated total population of the wards of three Counties we compared the estimated values with the Gold Standard values. The Gold Standard estimate was based on the total non-response in a LA. For example, say, in a LA the number of non-responses were 60 and there were only 2 wards. So the number of non-responses was allocated to each ward in a LA on the basis of:

1/2(%unemployed + %imputed)
For Ward 1, 1/2(%unemployed + %imputed)= 60% say
For Ward 2, 1/2(%unemployed + %imputed) = 40% say

This means out of 60 non-responses 36 persons were allocated to Ward 1 and 24 persons to Ward 2. Therefore, the total persons estimated for
Ward 1 = Enumerated + 36 = P1 and
Ward 2 = Enumerated + 24 = P2 and is known as Gold Standard estimate.

6.8 Results

We give the estimates for three counties, Inner London, Hampshire, and Wiltshire at Ward level with the 1991 census enumeration and the Gold Standard estimate in the summary Table 6.7, Table 6.8, and Table 6.9 respectively which are based on Model 1, Model 2, and Model 3 respectively. In case of Model 1 (Table 6.7) the Gold Standard estimates are always greater than the 1991 census count and the estimated values in Inner London. Like the Gold Standard estimate, in this area estimated values are always greater than the census count but lower than the Gold Standard estimate except in one case of Cripplegate Ward of the City of London local authority. In Hampshire the Gold Standard estimate is lower than the estimated value in two cases and equal to census counts in three cases and greater in all other cases. In this county estimated values are lower than the census count in three wards and equal in the same number of wards. In Wiltshire, the Gold Standard estimate, estimated value and the census counts are exactly the same except in one ward. The reasons for this may be that in this county the populations of the selected wards are small and the undercount or overcount rate is also small.

In case of Model 2 (Table 6.8) in Inner London area all the values of Gold Standard estimate are higher than the enumerated and the estimated values except in one where the estimated and the Gold Standard estimates are same (ward Abbey). In the Hampshire only in Bargate Ward of Southampton local authority Gold Standard estimate is lower than the estimated value and in Church Crookham Ward of Hort local authority Gold Standard estimate and the enumerated value are equal. In all other wards the Gold Standard estimates are higher than the estimated and enumerated values. Similar results were also observed in Wiltshire.

Comparison between estimated and census counts in the Inner London area show that only in the Crippligate Ward of the City of London local authority is the estimated value lower than the counted value and in all other cases it is higher than the counted values. In Hampshire results are different. Here, in half of the cases the estimated values are less than the enumerated values.

In the case of Model 3 (Table 6.9) one can observe almost the same result for Gold Standard estimates as that of Model 1, that is, Gold Standard estimated values are always higher than the enumerated and estimates values except in a few cases. In two wards of the Inner London area Gold Standard estimates are lower than the estimated values while in Wiltshire they are lower than the enumerated population in the same number of wards. In Hampshire only in one case is the Gold Standard estimate lower than the estimated value and in one other case it is same as that of enumerated population. Comparison between enumerated and estimated values shows almost identical results as those of Model 2. In Inner London only two estimated values are less than the enumerated values while the rest of the values are higher than the enumerated values. In the other two counties the results are the reverse. Only in two wards of Hampshire are the estimated values higher than the enumerated values while in Wiltshire all the enumerated values are higher than the estimated values.

Figure 6.1, 6.2 and 6.3 are the scatter plots for Model 1, Model 2 and Model 3 respectively. In all cases there are clear linear relationship between enumerated and estimated values. Figure 6.1c which is the scatter plot between estimated and Gold Standard estimate (Model 1) shows that only three values deviate from the linear relationship which implies that in these three cases deviation between the estimated undercount rate by the two methods are substantial. In all other cases the estimated undercount rates by the two methods are similar. One can observe similar results from Figure 6.2c and 6.3c for Model 2 and Model 3 respectively.

6.9 Conclusion

Among the three fitted regression models, Model 1, where undercount rates were estimated from the sex ratio for age group 20-24, selected only seven independent variables from the set of 35 independent variables, while Model 2 and Model 3 selected eight and ten independent variables respectively. The residual mean square of the Model 2 is only 0.583 which is much less than other two models. Model 2 also explains the maximum variation among the three models. From the correlation matrices we observed that no variables of Model 2 were highly or very poorly correlated with any other variables in the model. However, this is not the case for other two models. In the case of Model 1 two variables X_3 and X_{30} are highly correlated with each other which may affect the regression estimate and therefore we exclude one of them from our model. In the case of Model 3 low correlations were observed among the independent variables, which may be good for fitting the model and one can expect better estimate from this model. However, considering all the above criteria it seems to me that Model 2 fitted the data well.

Table 6.10: Statistics of the Models

Models	Standard Error	Residual Mean Square	R^2
Model 1	1.30	1.69	.329
Model 2	0.76	0.58	.401
Model 3	1.92	3.70	.297

6.10 Discussion

The method of combining regression and sample estimates and using the resulting estimates not only for estimates at the geographical, city, or Local Authority level, but as the basis for adjustment for all other small areas, such as Wards, is in need of careful investigation. Before the application of the method some critical questions must be resolved, at least in part.

1. There are several assumptions of the model. Are they met? The CVS has different samples for estimating the undercount rate of the 1991 census enumeration. In some cases it was very difficult to collect information from the selected samples. On the other hand, matching huge data files is a complex and erroneous process, and imputations for missing data leave plenty of room for error as well. It is therefore, likely that at least some bias is present in the estimates. It is extremely unlikely that the errors in the CVS estimates are not correlated over areas.

2. The values of K1, K2, and K3 depend on the sex ratio of the area concerned, the expected sex ratio of the same area, and the undercount rate of the male and female population. As we used the national undercount rate of the male and female population it remained constant for different local authorities. Therefore, practically the value of the Ks depends only on the sex ratio and the estimated expected sex ratio of the particular age group for the particular local authority. We carefully investigated all the estimated Ks of the 403 local areas and tried to explain why some the Ks have higher values. We excluded the unresolved cases from our estimates.

3. The model fitted at the Local Authority levels is then assumed to apply to Ward levels. This is questionable. Outliers effect regression estimates in very damaging ways (Freedman and Navidi, 1986). Moreover, the accuracy of the estimates for Wards may be far less than for those of Local Authorities.

Table 6.7: Estimate From Model 1

LA	Ward	Enu. Pop	K1	Gold Est
Inner London				
City of London	Cripplegale	035	032	037
Camden	Adelaid	214	224	225
Hackney	Brownswood	282	292	298
Hamm + Fulham	Addison	303	313	318
Haringey	Alexandra	169	170	172
Islington	Barnsbury	234	239	243
Kens + Chelsea	Abingdon	363	380	385
Lambeth	Angell	428	436	457
Lewisham	Bellingham	174	175	180
Newham	Beckton	165	167	170
Southwark	Abbey	190	196	197
Towerhamlet	Blackwall	098	100	102
Wandsworth	Balham	357	358	374
Westminister	Baker St.	222	241	251
Hampshire				
Basingstoke	Basing	294	289	294
East Hampshire	Alton Holybourne	031	032	034
Eastleigh	Bishopstoke	115	114	116
Fareham	Fareham East	071	071	072
Gosport	Alverstoke	597	601	621
Hort Church	Crookham	061	061	061
Havant	Barncroft	089	089	090
New Forest	Barton	024	025	024
Portsmouth	Charles Dickens	308	315	356
Rushmoor	Alxandra	174	176	180
Southampton	Bargate	584	604	680
Test Vally	Abbey	043	044	043
Winchester	Badger Farm	048	047	049
Wiltshire				
Kennet	Aldbourne	019	019	019
North Wiltshire	Allington	021	021	021
Salisbury	Alderbury	035	035	035
Thamesdown	Blunsdon	048	048	048
West Wiltshire	Adcroft	077	078	078

Table 6.8: Estimate From Model 2

LA	Ward	Enu. Pop	K1	Gold Est
Inner London				
City of London	Cripplegale	076	070	081
Camden	Adelaid	372	382	392
Hackney	Brownswood	584	603	616
Hamm + Fulham	Addison	662	680	697
Haringey	Alexandra	422	426	430
Islington	Barnsbury	507	519	527
Kens + Chelsea	Abingdon	416	422	441
Lambeth	Angell	819	839	874
Lewisham	Bellingham	360	361	373
Newham	Beckton	251	255	259
Southwark	Abbey	396	412	412
Towerhamlet	Blackwall	200	207	209
Wandsworth	Balham	863	865	904
Westminister	Baker St.	289	300	327
Hampshire				
Basingstoke	Basing	738	736	739
East Hampshire	Alton Holybourne	063	063	063
Eastleigh	Bishopstoke	325	320	329
Fareham	Fareham East	139	138	140
Gosport	Alverstoke	809	806	842
Hort Church	Crookham	199	198	199
Havant	Barncroft	200	197	202
New Forest	Barton	054	054	055
Portsmouth	Charles Dickens	502	519	581
Rushmoor	Alxandra	315	315	326
Southampton	Bargate	664	677	774
Test Vally	Abbey	091	091	092
Winchester	Badger Farm	184	182	188
Wiltshire				
Kennet	Aldbourne	046	046	046
North Wiltshire	Allington	060	060	061
Salisbury	Alderbury	074	073	073
Thamesdown	Blunsdon	066	065	066
West Wiltshire	Adcroft	151	151	154

Table 6.9: Estimate From Model 3

LA	Ward	Enu. Pop	K1	Gold Est
Inner London				
City of London	Cripplegale	112	104	119
Camden	Adelaid	748	757	788
Hackney	Brownswood	761	803	803
Hamm + Fulham	Addison	729	752	767
Haringey	Alexandra	878	912	895
Islington	Barnsbury	917	933	953
Kens + Chelsea	Abingdon	694	697	736
Lambeth	Angell	1073	1101	1145
Lewisham	Bellingham	586	581	607
Newham	Beckton	317	322	327
Southwark	Abbey	560	585	582
Towerhamlet	Blackwall	370	370	386
Wandsworth	Balham	1272	1265	1332
Westminister	Baker St.	329	345	373
Hampshire				
Basingstoke	Basing	1371	1380	1373
East Hampshire	Alton Holybourne	137	125	151
Eastleigh	Bishopstoke	858	837	869
Fareham	Fareham East	430	428	433
Gosport	Alverstoke	963	939	1002
Hort Church	Crookham	454	434	454
Havant	Barncroft	427	410	431
New Forest	Barton	106	099	107
Portsmouth	Charles Dickens	724	726	838
Rushmoor	Alxandra	444	438	459
Southampton	Bargate	872	857	1017
Test Vally	Abbey	189	181	191
Winchester	Badger Farm	343	330	351
Wiltshire				
Kennet	Aldbourne	120	117	119
North Wiltshire	Allington	144	137	146
Salisbury	Alderbury	203	198	201
Thamesdown	Blunsdon	167	158	167
West Wiltshire	Adcroft	240	225	244

Here:

Column 1 is the name of the local authority (LA).

Column 2 is the name of the ward of the Column 1 local authority.

Column 3 is the census enumerated population total of the ward.

Column 4 is the estimated ward population from the regression Models.

In the regression model 1, K1 was used as the dependent variable for the age group 20-24.

K2 is the estimated ward population from the regression model 2.

In the regression model 2, K2 was used as the dependent variable for the age group 25-29.

K3 is the estimated ward population from the regression model 3.

In this regression model K3 was used as the dependent variable for the age group 30-34.

Column 5 is the Gold Standard estimate of the ward population.

Chapter 7

Dealing With Missing Data

7.1 Introduction

In many censuses and sample surveys, especially those that involve human populations, there will be some units selected into the sample for which all or part of the survey data items are not obtained. The problems created from incomplete data are not only that they reduce the sample size which means less efficient estimates but also that standard complete-data methods cannot be immediately used to analyze the data. Moreover, there will necessarily be some bias in the survey estimates. There is no method of correcting nonresponse bias since the missing survey characteristics of the nonrespondents are, by definition, not available. However, it is believed that there are usually systematic differences between respondents and nonrespondents and no statistical technique can be relied upon to adjust all the differences. In order to hold this bias to a minimum level, the ideal way is to obtain complete data. However, in practice, it is impossible to complete the survey. Therefore, in the language of Dempster and Rubin (1987), "it is both desirable to minimize nonresponse by design and necessary to adjust for the residual incomplete data by analysis, recognizing that no adjustment can fully compensate for the missing data".

7.1.1 Types of Nonobservation

Nonobservation in censuses or in sample surveys occurs in three ways: noncoverage, total or unit nonresponse and item nonresponse.

Noncoverage represents a failure to include some of the target population in the sample frame; in consequence, they have no chance of appearing in the sample.

Noncoverage refers to the negative errors due to the exclusion of the elements that would properly belong in the sample. Positive errors of noncoverage also occur when some elements are included in the sample that do not belong there. The term gross coverage error refers to the sum of the absolute values of noncoverage and overcoverage error rates. The net noncoverage refers to the excess of noncoverage over overcoverage, and is their algebraic sum. In most social surveys it is much more difficult to measure the noncoverage, though it is a more common problem than overcoverage. In that sense net noncoverage is an acceptable measure of coverage problems (Kish, 1965).

Unit or total nonresponse occurs when none of the variables is measured for a unit or subunit. This may be caused due to

1. Not at home
2. Refusal
3. Incapacity or inability
4. Not found or
5. Lost schedules (Kish, 1965).

Item nonresponse occurs when most of the questions for units are answered, but for certain questions either no answer is given or the answer is judged to contain a gross error and is deleted during editing. Item nonresponse may arise due to any of the following cases:

1. Lack of records
2. Failure of memory
3. Accidental mistakes in responding
4. Unwillingness to give the right answer
5. Illegible entries
6. Refusal to give the right answer
7. Wrong answer arising from pride, called prestige-bias, etc..

7.1.2 Effect of Nonobservation

In the study of nonresponse it is helpful to think of the survey population as composed of two groups, respondents and nonrespondents. Cochran (1977) defined this division into two distinct strata as an oversimplification. However, for Illustration purposes this simple model suffices.

Suppose that the aim of the survey is to estimate $\bar{Y}$, the population mean. For simplicity we will consider a simple random sample. Let N denote the population size and N_1 and N_2 the number of respondents and nonrespondents respectively such that $N_1 + N_2 = N$. Let $W_1 = N_1/N$ and $W_2 = N_2/N$ denote the proportion of respondents and nonrespondents ($W_1 + W_2 =1$), and $\bar{Y}_1$ and $\bar{Y}_2$ the means of the survey variable for the groups. Under the conditions of the survey the mean is

$$\bar{Y} = W_1 \bar{Y}_1 + W_2 \bar{Y}_2.$$

Since the survey fails to collect data for the nonrespondents, it produces the estimate $\bar{Y}_1$. The use of a mean response $\bar{Y}_1$ to estimate the mean $\bar{Y}$ causes a bias, which is

$$\bar{Y}_1 - \bar{Y} = \bar{Y}_1 - (W_1 \bar{Y}_1 + W_2 \bar{Y}_2) = \bar{Y}_1(1 - W_1) - W_2 \bar{Y}_2 = W_2(\bar{Y}_1 - \bar{Y}_2) \tag{7.1}$$

This bias is seen to depend on two factors, W_2, the proportion of nonresponse and $(\bar{Y}_1 - \bar{Y}_2)$, the difference between the means for respondents and nonrespondents. If the proportion of nonresponse W_2 and the difference of $(\bar{Y}_1 - \bar{Y}_2)$ are both small, the bias should be negligible. It is therefore important to keep the nonresponse sufficiently small to guarantee that when $(\bar{Y}_1 - \bar{Y}_2)$ is multiplied by W_2 the result will not be large.

The population total Y can be estimated in two different ways. The first is to use the simple inflation estimator F_y, whose expected value is $N_1 \bar{Y}_1 = NW_1 \bar{Y}_1$. The inflation factor F is the inverse of the sampling fraction f, an unbiased estimator of Y. Bias due to using $N_1 \bar{Y}_1$ as an estimator of Y is thus

$$B(\hat{Y}) = Y_1 - Y = -Y_2 = -N_2 \bar{Y}_2 \tag{7.2}$$

which depends on the size of the nonresponse group and its mean. This bias will be small if the total for the nonrespondents is small. The alternative is to use $N\hat{Y}_1$, which has a bias of $N_2(\bar{Y}_1 - \bar{Y}_2)$. We already mentioned the two factors on which this bias depends. Here it is equivalent to assuming that the means of the two groups of respondents and nonrespondents are equal.

When comparing the means of two subclasses, the difference $(\bar{Y}_a - \bar{Y}_b)$ has the bias

$$(\bar{Y}_1 - \bar{Y})_a - (\bar{Y}_1 - \bar{Y})_b = [W_2(\bar{Y}_1 - \bar{Y}_2)]_a - [W_2(\bar{Y}_1 - \bar{Y}_2)]_b \tag{7.3}$$

In most cases, though not always, the biases in the individual groups may tend to cancel. This cancelling of biases may not occur if the effect of nonresponse differs from one class to another and if the separate effects of nonresponse are different in different classes.

7.1.3 Compensation Procedure for Missing Data:

Two strategies for handling missing data in surveys are common in practice, namely weighting adjustments and imputation techniques. The U.S. Bureau of Census used both of these two types of procedures in combination to handle missing data, when it evaluated the population census of 1980. Imputation techniques assign values for missing responses in various sub-groups of the sample in order to compensate for the sub-groups' differing response rates, while weighting adjustments increase the weight of the respondents. The choice between these two types of procedure for handling a particular type of missing data depends mainly on two things. First, the amount of information available on the unit and second, how the missing data arose, that is, whether from noncoverage, unit nonresponse or item nonresponse. Generally, compensation for noncoverage and unit nonresponse are made by weighting adjustments while item nonresponse is treated by imputation. "In practice there is no reason why unit nonresponse could not be handled by imputing all the survey items for missing units. Conversely, item nonresponse could be handled by assigning a set of weights to the respondents of each partly recorded variable, although the resulting proliferation of weight may create difficulties for analysis", (David et al 1983).

Besides the above two reasons of selecting weighting adjustments and imputation methods, there are another three considerations involved in making an ppropriate choice of compensation procedure. These considerations are :

1. the precision of the resulting estimates;
2. the estimation of standard errors of the estimates; and
3. the suitability of the compensated data set for producing estimates for a variety of different parameters.

7.1.4 Imputation:

Imputation is one way to deal with missing data. Here we will briefly discuss some of the general imputation methods used in practice:

Mean Imputation:

There are different types of mean imputation methods. Among them some of the methods used are:

a) Mean imputation overall (MO)
b) Random imputation overall (RO)
c) Mean imputation within classes (MC), and
d) Random imputation within classes (RC)

a) Mean imputation overall (MO): In this method the mean value from the respondents is assigned for the missing value.

b) Random imputation overall: In this method each of the missing values is assigned a value randomly selected from the respondents.

c) Mean imputation within classes: In this method total samples are divided into imputation classes according to some criteria. With each class, the respondents mean of that class is assigned as an imputed value for all the nonrespondents in that class.

d) Random imputation within classes: In this method a randomly selected value from the same class is assigned for the missing value for that particular class.

Mean imputation methods are simple to use. However, the methods have, in general, some undesirable properties. First, as the sample size is reduced due to nonresponse, the standard variance formula will systematically underestimate the true variance. Second, estimates of quantities that are not linear in the data, such as the correlation between a pair of variables, cannot be estimated consistently using complete-data methods on the completed data. Third, the empirical distribution of the sample values is distorted by the imputed means, which is important when studying the shape of the distribution (Little and Rubin, 1987).

Hot Deck Imputation:

In the hot-deck method of imputation a value from the respondents is duplicated to assign for the missing value. In this procedure, all the sample units are classified into disjoint groups so that the units are as homogeneous as possible within each group. A reported value is imputed for each missing value which is in the same classification group. Thus, the assumption is made that the nonrespondents follow the same distribution as the respondents within each classification group. In the sequential hot-deck procedure the sample is put in some type of order within each classification group. Each missing value is replaced by the recorded value in the same classification group. A major attraction of this procedure is its computing economy. In general, hot-deck procedures have three advantages. They reduce the nonresponse bias, produce complete data sets, and preserve the distribution of the population as represented by a sample (Ford, 1983, Vol.-II).

Cold Deck Imputation:

In cold-deck imputation, the missing value is replaced by a constant value using information from data other than the current sample. For example, values of relations from previous surveys may be used to impute value for missing values.

Regression and Stochastic Regression Imputation:

In the regression imputation procedure, a missing item is replaced by predicted values from a regression of the missing item on an item observed for the unit, usually calculated from units with both observed and missing variables present.

In the stochastic regression imputation procedure, the missing value is replaced by a value predicted by regression imputation plus some type of randomly chosen residual.

With the help of careful development of the regression model, regression imputation has the potential to produced imputed values closer to the true values. However, the construction and assessment of a good regression model is a time consuming operation, and it seems unrealistic to consider its application for all the items with missing values in a survey. Attention also needs to be given, in using regression imputation, to problems of estimating several missing items on the same record (Kalton, 1983).

Deductive Imputation:

In the deductive imputation procedure, the missing value is imputed in situations in which a missing response can be deduced with certainty, or with high probability, from other information on the record. For example, if a respondent's sex is missing, but the person has a male name and is known to be married to a female, the sex of the respondent may be deduced to be male. In a panel survey with a variable that remains almost constant over time, a missing value may be assigned from the record value.

The deduction method of imputation essentially depends on some redundancy in the information collected so that edit constraints can determine the missing values on some items (Kalton, 1983).

Flexible Matching Imputation:

Flexible matching imputation is a modified hot-deck procedure that has been used by the Bureau of the Census Income Supplement of the CPS since 1976. The procedure of the method starts by sorting respondents and nonrespondents into a large number of classes, constructed from a detailed categorization of a sizeable set of auxiliary variables. The matching of the nonrespondents with the respondents is done on a hierarchical basis in the sense that if a nonrespondent cannot be matched with a respondent in the imputation class, classes are collapsed and the match is made at a lower level. Three hierarchical levels are defined for this purpose, with the lowest level being such that a match can always be made. This flexible matching procedure enables closer matches to be secured for many nonrespondents than does the traditional hot-deck procedure. The procedure also avoids the multiple use of respondents in classes where the number of nonrespondents does not exceed the number of respondents.

Substitution:

In the substitution method, unit nonresponses are dealt with within the field-work stage of the survey by replacing nonrespondents by alternative units not selected into the sample rather than imputing data from respondents or adjusting the weights of the respondents. In designing a substitution procedure, arrangements are made in such a way that the substitutes have similar survey characteristics to those of the nonrespondents. In general, two basic types of substitution procedure are used:

a) selection of a random substitute, and
b) selection of a special designated substitute.

In a random substitute procedure, to replace each nonrespondent, an additional population unit is selected on a probability basis. Usually, to have a more similar type of substitution for the nonrespondent, the substitute is chosen from a restricted population subgroup (e.g., the same block, area, strata, or group of strata from which the nonrespondent was selected). To avoid any delay and trouble that would be involved in selecting a substitute for a nonresponding sample unit after the data collection activities have begun, many random substitution procedures select potential substitutes prior to the data collection phase of the survey.

In a specially designated substitute procedure, one or more backup units are identified for substitution. These identified units have similar characteristics to those of the nonrespondents (e.g., a geographic neighbour of a nonrespondent or a unit that has specified characteristics identical with or similar to those of the sample unit.

The main advantage of the use of the substitution is that it increases the survey sample size, and therefore reduces the variances of survey estimates. The other advantage is that the sample will be balanced with respect to sample size per substitution class. This balance has certain practical advantages.

Among the disadvantages, the first disadvantage to the use of substitution is that, sometimes, an interviewer and perhaps a research analyst, may view a backup unit as one that is just as good (or nearly as good) as the unit initially selected. As a result, effort to obtain response from originally selected units may not be as intense as it would if no substitution were available. The second disadvantage of the use of substitution is that, when the survey response rate is estimated, there is a tendency to ignore the level of substitution used. This will, of course, overestimate the survey response rate and underestimate the nonresponse bias.

7.1.5 Multiple Imputation:

In multiple imputation methods each missing value is replaced by a vector of $M \geq 2$ imputed values; this idea was originally proposed by Rubin (1978), although the idea appears in Rubin (1977). The M values are ordered in the sense that the first components of the vectors when substituted for the missing value result in one data set, the second components result in a second data set, and so on. To analyze each data set, standard complete-data methods are used. When the M sets of imputations are repeated random draws under one model of nonresponse, the M complete-data inferences can be combined to form one inference that properly reflects uncertainty due to nonresponse under that model.

Multiple imputation shares all the basic advantages, namely, the ability to use complete-data methods of analysis and the ability to incorporate the data collector's knowledge, with single imputation methods. Besides these, there exist three more extremely important advantages to multiple imputation over simple imputation. First, when imputations are randomly drawn in an attempt to represent the distribution of the data, multiple imputation increases the efficiency of estimation. Second, when the multiple imputation represents repeated random draws under a model for nonresponse, valid inferences --that is, ones that reflect the additional variability due to the missing values under that model --are obtained simply by combining complete-data inferences in a straightforward manner. Third, by generating repeated randomly drawn imputation under more than one model, it allows the straightforward study of the sensitivity of inferences to various models for nonresponse simple using complete-data methods repeatedly.

There are also three obvious disadvantages of multiple imputations relative to simple imputation. First, more work is needed to produce multiple imputations than single imputations. Second, more space is needed to store a multiply-imputed data set. Third, more work is also needed to analyze a multiply-imputed data set than a singly-imputed data set. These disadvantages are not serious when M is modest.

7.1.6 Proposed Method of Imputation

7.1.7 Introduction

In section 7, from 7.1 to 7.9 we have described different methods that are used to deal with missing data. In this part we will discuss only our proposed method of imputation. It is important to mention that our main purpose is to estimate the undercount rate in the census, for the whole nation as well as for different sex, race, and age-groups, by applying the dual system method of estimation with the help of the six different samples, which we mentioned in the previous chapter. Therefore, our problems arising from missing units, and here we will deal only with missing units rather than missing items which can be observed in two different ways, viz:

A. During the investigation of the six samples by the CVS interviewers and
B. During the estimation of the total population by the dual system methods of estimation.

7.1.8 Missing units from the six CVS samples

During the investigation of the six samples by the CVS interviewers case by case, it is expected that they will find some missing households or persons. When CVS interviewers find say, a person from any of the above six samples, that person will be weighted by the inverse of the probability of selection to compensate for unequal probability of selection. That is, our ultimate figure of missing persons or units is the outcome of mathematical procedure, which means, we will have no idea about some of the missing persons or units, except the geographic position. We therefore have four different types of missing units which are:

a) missing houses with some information.
b) missing houses with no information.
c) missing persons with some information.
d) missing persons with no information.

For the imputation of the above four types of missing units we will construct eight different Tables taking data from all the selected ED blocks as well as from the last census count. The first Table will be constructed according to the types of houses, that is, how many people live in a house. Within an ED block there may be different types of houses. In the following example we assume that we have only six types of houses and in the sample there are 100 houses only. The Table 7.1 is as follows:

Table 7.1: Types of Houses and Distribution of Persons

Type of Houses	No. Of Person live in the house	Frequencies	Probabilities
01	01	12	001-012
02	02	19	013-031
03	03	24	032-055
04	04	32	056-087
05	05	08	088-095
06	06	05	096-100

The second Table (Table 7.2) will be constructed by taking sample units from the selected ED blocks and from the last census count. These sample units will be classified into disjoint groups according to age-groups so that the units are as homogeneous as possible within each group. In each group there will be a large number of rows where each row will define each person. The rest of the six Tables (Table 7.1.1; Table 7.1.2 etc.) will be constructed according to the types of houses. For each type of house one Table will be constructed, taking data from all the selected ED blocks and from the last census count. In each Table there will be a large number of rows and each row will be divided into 2 to 6 sub-rows where each row will define each housing unit and each sub-rows will define each person with alternative set of variables.

Appropriate weight must be given for the different age groups, sex and race in fixing the number of alternative set of variables.

Table 7.2: Distribution of Persons According to Age-Groups

S. No.	Male/Female	Age group	Race
Group 1		Age-group, 0-16	
001	Male	14	White
002	Male	12	White
003	Female	15	White
..	..	..	..
..	..	..	..
$00n_1$	Male	11	Black
Group 2		Age-group, 17-44	
001	Male	42	White
002	Male	22	White
003	Female	39	White
004	Male	32	Black
..	..	..	..
..	..	..	..
$00n_2$	Male	26	White
Group 3		Age-group, 45-64	
001	Male	54	White
002	Male	62	White
003	Female	49	White
004	Male	63	Black
..	..	..	..
..	..	..	..
$00n_3$	Male	62	White
Group 4		Age-group, 65+	
001	Male	74	White
002	Male	82	White
003	Female	94	White
004	Male	73	Black
..	..	..	..
..	..	..	..
$00n_2$	Male	67	White

After constructing the Tables we will assign values for missing units for each of the above categories independently in the following ways.

a) Missing houses with information:

When we have information about the missing house or houses we will utilize that information for assigning values for the missing unit. Say, for example we know that missing housing unit is of type two of Table 7.1. That is, two persons were in the missing unit. If we have no other information then we will select randomly one row from the Table 7.1.2 which consists of two subrows, "a and b" and will assign values from these two subrows for the missing persons. On the other hand if we have further information about some of the variables of the missing units, then we will match that information with the values of each row of Table 7.2 and will assign the values of that row which match more closely with the information of the missing unit.

b) Missing houses with no information:

When we have no information about the missing unit we will assign values in two steps. In the first step we will decide type of house, and in the next step the row of the corresponding Table. For example, say one house is found missed from the census count and we have no information about that unit. From Table 7.1 we will decide the type of housing unit missed by the census and say, the selected type is 3. It means that in the missing house there were three persons and we have to impute variables for these three persons. For this we will select a row from Table 7.1.3 which consists of three subrows and impute values from these three subrows taking values of one subrow for one missing person.

Table 7.1.1 Single person live in each house

S. No.	No. of person live	Male/Female	Age-group	Race
001	01	Male	45	White
002	01	Male	52	White
003	01	Female	49	White
004	01	Male	32	Blake
..		..		
..		..		
150	01	Male	26	White

c) Missing person with information:

When we have information about some of the variables of the missing unit, we will match that information with the variables of each row of Table 7.2 and will assign the values of that row which match more closely for the missing unit.

d) Missing person with no information:

When we have a missing person with no information we will impute variables for this missing unit in several steps. In the first step we will select type of house and say our selected type of house is 4, that is, the missing person is from a house were 4 persons were lived. In the second step we will select a row from the corresponding Table 7.1.4. In the third step we will select a subrow from this selected row and will impute the variables for the missing unit.

Table 7.1.2 Two persons live in each house

S. No.	No. of person live		Male/Female	Age-group	Race
001	02	a	Male	42	White
		b	Female	36	White
002	02	a	Male	62	White
		b	Female	59	Blake
003	02	a	Male	32	White
		b	Male	30	White
..			..		
..			..		
150	02	a	Female	48	White
		b	Female	08	White

We will construct Table 7.1.3, 7.1.4, 7.1.5 and 7.1.6 (Appendix 4) for the house of types 3, 4, 5 and 6 respectively following the same procedure of Table 7.1.1 and 7.1.2 above.

7.1.9 Missing units from the dual system estimation

After estimating the total population by the dual system method of estimation we will estimate the missing persons who were missed by the both methods i.e. by the census enumerators as well as by the CVS investigators. By the same method we will also estimate the total number of missing persons by sex, race and age-groups. Therefore, by applying the dual system method of estimation we will estimate

a) Total number of missing persons.
b) Total number of missing persons according to the race.
c) Total number of missing persons according to the sex.
d) Total number of missing persons according to the age-groups.

Now our problem is to distribute these estimated persons according to their race, sex and age-groups. We will solve this problem for each of the selected ED block in the following ways.

1. Say, for example, in a selected ED block, it is estimated that 19 persons are missing. The distribution of these missing persons could be as table 7.5 (arbitrarily).

Table 7.5: Distribution of Missing Persons

a	Black	11
	Non-black	08
b	Male	11
	Female	08
	Age-group	
	0-16	05
	17-44	06
	45-65	04
	65+	04

2. In the following table (Table 7.6) we distribute the missing persons according to the information of the Table 7.5 by an iterative method with initial conditions determined by the marginal distribution and the observed association in the census data. Here, we will also utilize the information, if any, available from the CVS investigation.

Table 7.6: Distribution of Missing Persons by Iterative Method

Age-group	Male		Female		Total
	Black	Non-black	Black	Non-black	
00-16	02	01	01	01	05
17-44	02	01	02	01	06
45-64	01	02	01	00	04
65+	01	01	01	01	04
Total	11		08		19

3. Now the question is how we will decide out of 19 (say) persons how many of them are black or non-black, male or female etc.. This will be done proportionally from the total number of missing persons for the whole country. One example is given below (Table 7.7).

Table 7.7: Proportional Distribution of Missing Persons

	Country	ED block
Total missing persons	100	10
Total male missing persons	58	06
Total female missing persons	42	04
Total black missing persons	30	03
Total non-black missing persons	70	07

4. After the distribution of the missing persons according to race, sex and age-groups we will match the information of the missing persons with each of the disjoint groups of Table 7.2. We will draw one row randomly from the groups which match more closely for the missing unit and assign the values of that row for the missing unit.

7.2 Conclusion

The main advantage of our method is that it generates a complete data set that may be readily used for many different forms of analysis. As the method is based on the current census counts of the ED blocks, it is expected that the method will produce sufficiently accurate results. The main drawback of imputation is that incorrect imputation may introduce an element of bias into the results.

One aspect of this is that, in the case of a missing house with no information, our method assigns values in two steps. In the first step we select the type of missing house randomly, that is, how many persons live in that missing house. For example, say our randomly selected type of house is three, that is three persons live in that missing house and hence in the second stage we impute variables for these three persons. This is may be wrong, because in fact in that missing house two persons may live and we therefore overestimate the population. Moreover in our method we need to construct too many tables which is time consuming, costly and may be also difficult.

Chapter 8

Assessing Error

Introduction

So far we have discussed different kinds of sampling and non-sampling errors and the errors that affect the population estimate when the dual system estimator is used to estimate the population total, by using all the six CVS samples (Visual list, Vacant, Absent, Non-Residential, Multi-Households and Quality Check sample) drawn from the ED workload with the census count. Among these six samples as listed in Chapter 3.3 we will in this Chapter call the first five samples a P-sample and the last one, that is, the Quality Check (also known as Census Co-operative) sample an E-sample. The P-sample is designed for the estimation of people missed by the census enumeration, and the E-sample is designed for the estimation of the number of erroneous enumerations. Here we will discuss the total error of our statistic and the method of estimating this error. In this chapter this statistic, the difference between the true value and the estimated value is defined as the total error. The main problem of estimating the total error is that we do not know the true population value. To overcome the problem Ericksen and Kadane (1985) suggested using a demographic estimate. We have already discussed the limitation of the demographic estimates. Mulry and Spencer (1991) suggested identifying all the sources of error, and estimating their magnitudes to make the estimate. To evaluate the error of our estimate we will follow the procedure suggested by Mulry and Spencer (1991). For that we divide the total error into three components with the assumption that all sources of errors will fall in any of the three components. We will try to estimate the magnitude of error of these three components independently and then the aggregate of these three components of error will give the total error. The different sources of errors we describe here are not a complete list of the sources of errors that may affect the population estimate. There may be many more. However, the advantage of the methodology described here is that it will give a guide to identify different sources of errors, and the methods of estimating their magnitude.

8.2 Background

To estimate the population total the application of the dual system estimator involves two incomplete lists of the population (Wolter, 1986). The first list is the census enumeration of the population not living in institutions or homeless. By definition this list excludes erroneous enumeration and imputations. The second is an implicit list of those persons covered by the sampling frame for the P sample, whom we call the P sample population. This list would be obtained if the P sample were conducted for the entire U.K. with no measurement error or missing data. The sampling frame itself is not a list of people, but of ED blocks. A person may be excluded from the P sample population for a number of reasons: the residence may be on a block that could be sampled but the residence would not be identified as an occupied dwelling unit, or the residence could be identified but the person would not be reported in the CVS.

It is assumed that whether the i'th individual in the population of size N is in the census or not and in the P sample or not are to be random events with positive probabilities as shown in Table 8.1 which also shows the true population size. Thus person i has probability $P_{i1+} = P_{i11} + P_{i12} > 0$ of being enumerated in the census. (The subscript + denotes summation over possible values of the subscript.)

Table 8.1: Prob. of Inclusion and Population Size in a Cell

Census	P-Sample		
	In	Out	Total
In	$P_{i11}\|x_{11}$	$P_{i12}\|x_{12}$	$P_{i1+}\|x_{1+}$
Out	$P_{i12}\|x_{21}$	$P_{i22}\|(x_{22})$	$P_{i1+}\|(x_{2+})$
Total	$P_{i+1}\|x_{+1}$	$P_{i+2}\|(x_{+2})$	$P_{i++}\|(x_{++})$

The total population size of the non-institutionalized, non-homeless population on Census Day is shown in Table 8.1 as $x_{++} = N_T$. The census count N_c is not the same as x_{1+}, the number of persons 'in' the original enumeration because erroneous enumerations are not included in x_{1+} or x_{2+} although they are included in N_c. Even if we could observe the x_{jk}'s in the first row and first column, the x_{jk}'s in parentheses would not be observed directly but would have to be estimated. A strategy for estimating N_T would be to divide the number of persons enumerated in the census, x_{1+} by an estimate from the CVS of the proportion of the population that were enumerated in the census, x_{11}/x_{+1}. The resulting estimator,

$$N^* = x_{1+}x_{+1}/x_{11}$$

is called the basic dual system estimate (DSE).

8.3 Empirical DSE

In applications of dual system estimation, in the CVS we cannot directly observe x_{1+}, x_{+1} and x_{11}, because the P sample is only a sample of the x_{+1} members of the P-sample population. Moreover, the problems that were present in the census such as missing data, uncertainties about ED boundaries, Census Day address inaccuracies, geocoding errors, and undetected fabrication were also present in the P-sample, thereby leading to as underestimation of x_{11}. To account partially for the underestimation of x_{11}, we will adjust the number of census enumerations, $\tilde{N}_C$, for erroneous and imputed enumerations and for people with insufficient information to allow a match in the following way:

$$\tilde{N}_{CE} = \tilde{N}_C - I_c - \tilde{I}_E - \check{E}_E$$

As $\check{E}_E$ is not known, we will substitute an estimate, $\hat{E}_E$, and obtain

$$\check{N}_{CE} = N_C - I_c - \tilde{I}_E - \hat{E}_E$$

where

$\tilde{N}_{CE}$ is the adjusted census count for erroneous and imputed enumerations.

$\check{N}_{CE}$ is the estimate of the adjusted census count $\tilde{N}_{CE}$.

I_C is the number of persons imputed into original enumeration.

I_E is the number of people counted in the census for whom names are not available.

$\tilde{I}_E$ is the weighted number of census enumerations (from E sample) with insufficient information for matching, and

$\hat{E}_E$ denotes the weighted number of erroneous enumerations that were included in the E sample.

$\check{E}_E$ is the estimate of $\hat{E}_E$.

Due to the Census Day address inaccuracies, missing data in the P sample, and many other known and unknown reasons we cannot directly observe N_P (also denoted by x_{+1}), but instead use an estimator, $\check{N}_P$. Similarly we will also estimate the size of x_{11} by using the estimator $\check{N}_{CP}$ as we cannot observe the value of x_{11} directly. Therefore the estimate of $\check{N}_T$ by using empirical DSE is

$$\check{N}_T = (\check{N}_P \check{N}_{CE}) / \check{N}_{CP}$$

Hence, the estimate of the percent net undercount, or net undercount rate in the enumeration is

$$\hat{U} = 100(\check{N}_T - \check{N}_C) / \check{N}_T \tag{8.1}$$

Notation

The subscript $_P$ refers to a quantity based on the P-sample;

The subscript $_C$ refers to a quantity based on the Census;

The subscript $_E$ refers to a quantity based on E-sample;

The subscript $_{CE}$ refers to a quantity based on Census + E-sample;

The subscript $_{CP}$ refers to a quantity based on Census + P-sample;

The symbol ˘ over any letter denotes an estimator subject to sampling and nonsampling error.

The symbol ˜ over any letter refers to a quantity subject to sampling error.

8.4 Total Error and Partitioning the Total Error

Let us start with the simple model

$$\check{N}_T - N_{True} = (\check{N}_T - \acute{N}) + (\acute{N} - N_{True}) \tag{8.2}$$

Where:

$\check{N}_T$ is the Empirical DSE of total population.

N_{True} is the true population, which is unknown.

$\acute{N}$ is the target population total, subject to sampling error, which is also unknown.

N^* is the Basic DSE of total population.

The first term refers to sampling error and the second term contains all non-sampling errors. The second term of the above equation can be split into two parts. Hence the above equation becomes:

$$\check{N}_T - N_{True} = (\check{N}_T - \acute{N}) + (\acute{N} - N^*) + (N^* - N_{True}) \tag{8.3}$$

The middle and the last terms of the right hand side of the equation are known as the measurement and model error respectively. Total error, therefore can be partitioned into three components, viz, sampling error, measurement error and model error. We believe that all kinds of errors in the estimation of census undercount rate will fall in one of the above categories. Of course, we can partition the total error into more components, such as mixed error, but there effect may be very little and can be negligible (Mulry and Spencer, 1991).

8.5 Sampling Error:

8.5.1 Sources of Error

In the 1991 census validation survey (CVS) six type of samples were used to estimate the census undercount rate in the U.K. In our proposed model we will use all these six samples, all of which were collected from the same sample block. As the cell frequencies x_{11}, x_{12}, and x_{21} of our model are based on the above six samples, they will be subject to sampling error.

8.5.2 Definition

If a different sample had been drawn, a different empirical DSE would have been observed. We will estimate this sampling error by applying the **Jackknife method for variance estimation** described in the following.

Let N denote a finite population of identifiable units. Attached to each unit in the population is the value of an estimation variable, say Y. Thus, Y_i is the value of the i-th unit with i = 1,..., N. We are interested to estimate the population total which is denoted by

$$Y = \Sigma \, Y_i \, ; \, i = 1.......N \tag{8.4}$$

Now suppose that a probability proportional to size with replacement (ppswr) sample of size n is selected from N using probabilities (p_i), with $\Sigma \, p_i = 1$ and $p_i > 0$ for i = 1,...N. The usual estimator of the population total Y and its variance are given by

$$\hat{Y} = 1/n \, \Sigma \, y_i / \, p_i \, ; \, i = 1,2..............n \tag{8.5}$$
And
$$Var(Y) = 1/n \, \Sigma \, p_i \, (Y_i/p_i - Y)^2 \, ; \, i = 1.......N \tag{8.6}$$

respectively. Let $\check{N} = \hat{Y}$ and let us partition the complete sample into k groups of m observations each, and for convenience assuming that n, m, and k are all integers and n = mk. Let $\check{N}_{(i)}$ be the estimator of the same functional form as $\check{N}$, but computed from the reduced sample of size m(k - 1) obtained by omitting the ith group, and define

$$\check{N}_i = k \, \check{N} - (k - 1) \, \check{N}_{(i)} \tag{8.7}$$

Quenouille's estimator of the total N is the mean of the $\check{N}_i$

$$\check{N}_Q = \Sigma \, \check{N}_i / \, k \, ; \, 1 = 1,2,.....k \tag{8.8}$$
and the $\check{N}_i$ are called `pseudovalues'. The Jackknife estimator of variance is
$$V \, (\check{N}) = 1/k(k-1) \, \Sigma \, (\check{N}_i - \check{N}_Q)^2 \tag{8.9}$$

Estimation Procedure of Jackknife Method

In our proposed method of estimation, we assume that the population total N is unobservable and to be estimated. We also assume that the census and CVS samples both fail to include at least some portion of the population total N and the union of the two lists also fails to include some portion of the population total N. After matching the census data with the CVS sample, we produce the following data :

Table 8.2: Matching Outcome

Census	CVS-Sample(PEP)		
	In	Out	Total
In	$\hat{e}_{11}$	$\hat{e}_{12}$	$\hat{e}_{1+}$
Out	$\hat{e}_{21}$		
Total	$\hat{e}_{+1}$		

And

Table 8.3: Matching Outcome of ith ED

Census	CVS-Sample(PEP)		
	In	Out	Total
In	$\hat{e}_{11i}$	$\hat{e}_{12i}$	$\hat{e}_{i1+}$
Out	$\hat{e}_{21i}$		
Total	$\hat{e}_{i+1}$		

for i=1,2,3.....n. The subscript `i' is used to denote an estimator prepared from the i-th ED block.

In our example (Table 8.4) we have four selected ED blocks. Therefore, we have:

$\hat{e}_{11} = 1/4\Sigma\ \hat{e}_{11i}$; 1= 1......4.

$\hat{e}_{12} = 1/4\Sigma\ \hat{e}_{12i}$

$\hat{e}_{21} = 1/4\Sigma\ \hat{e}_{21i}$

$\hat{e}_{1+} = 1/4\Sigma\ \hat{e}_{i1+}$

$\hat{e}_{+1} = 1/4\Sigma\ \hat{e}_{i+1}$

The dual system estimator $\check{N}^*$ for this example is:

$$\check{N}^* = \hat{e}_{1+}\, \hat{e}_{+1}\, /\, \hat{e}_{11}$$

and the estimator obtained by deleting the ith block is:

$$\check{N}^*_{(i)} = \hat{e}_{(i)1+}\, \hat{e}_{(i)+1}\, /\, \hat{e}_{11(i)}$$

where

$$\hat{e}_{11(i)} = 1/3 \sum \hat{e}_{11i'}\; ;\; i' = 1,......3 \text{ and } i' \neq i$$

$$\hat{e}_{(i)1+} = 1/3 \sum \hat{e}_{i'1+}\; ;\; i' = 1,......3 \text{ and } i' \neq i$$

$$\hat{e}_{(i)+1} = 1/3 \sum \hat{e}_{i'+1}\; ;\; i' = 1,......3 \text{ and } i' \neq i$$

The pseudovalues are defined by:

$$\check{N}_i = 4\,\check{N}^* - 3\,\check{N}^*_{(i)}$$

Quenouille's estimator by:

$$\check{N}_Q = 1/4 \sum \check{N}_{(i)}\; ;\; i = 1,........4$$

and the Jackknife estimator of variance by:

$$V_1\,(\check{N}) = 1/4(3) \sum (\check{N}_i - \check{N}_Q)^2\; ;\; i = 1,....4$$

The conservative estimator of variance is:

$$V_2\,(\check{N}) = 1/4(3) \sum (\check{N}_i - \check{N}^*)^2\; ;\; i = 1,.....4$$

8.5.3 Example

The values of different columns of the Table 8.4 were taken arbitrarily just to illustrate the Jackknife method of estimation.

Table 8.4. Hypothetical Data

	$\hat{e}_{11i}$	$\hat{e}_{12i}$	$\hat{e}_{21i}$	$\hat{e}_{i1+}$	$\hat{e}_{i+1}$
$ED_{1.3}$	73	03	20	76	93
$ED_{5.2}$	76	05	20	81	96
$ED_{6.3}$	70	06	16	76	86
$ED_{9.4}$	74	06	15	80	89

From Table 8.4 we have :

$\hat{e}_{11} = 1/4 \Sigma\ \hat{e}_{11i} = 74.$

$\hat{e}_{12} = 1/4 \Sigma\ \hat{e}_{12i} = 05$

$\hat{e}_{21} = 1/4 \Sigma\ \hat{e}_{21i} = 18$

$\hat{e}_{1+} = 1/4 \Sigma\ \hat{e}_{i1+} = 78$

$\hat{e}_{+1} = 1/4 \Sigma\ \hat{e}_{i+1} = 91$

Therefore, the dual system estimator for $\check{N}^*$ for this example is

$\check{N}^* = \hat{e}_{1+} \hat{e}_{+1} / \hat{e}_{11} = (78)(91)/74 = 96$

And the estimates obtained by deleting the ith block are given in Table 8.5.

Table 8.5. Estimates Obtained by Deleting ith Block

i	$\hat{e}_{11i}$	$\hat{e}_{1+}$	$\hat{e}_{+1}$	$\check{N}^*_{(i)}$	$\check{N}_i$	$(\check{N}_i - \check{N}^*_i)$	$(\check{N}_i - \check{N}^*_{(i)})$
1	73	79	90	97	93	01	16
2	72	77	89	95	99	49	16
3	74	79	93	99	87	25	144
4	73	78	92	98	90	04	64

Therefore, Quenouille's estimate is:

$\check{N}_Q = 1/4 \Sigma\ \check{N}_{(i)} = 92$

and the Jackknife estimate of variance is

$V_1 (\check{N}) = 1/4(3) \Sigma\ (\check{N}_i - \check{N}_{Qi})^2 = 11.25$ and S.E. = 3.35

The conservative estimator of variance is:

$$V_2 (\check{N}) = 1/4(3) \Sigma (\check{N}_i - \check{N}^*)^2 = 20.00 \text{ and S.E.} = 4.47$$

8.5.4 Jackknife Method in the Case of a Stratified Sample

For the application of the Jackknife method in case of stratified sampling let us assume that the population is divided into $L=4$ strata (Table 8.6), where N_h describes the size of the h-th stratum. Suppose sampling is carried out using pps with replacement within the strata in the same way as before. However, the population sizes in this example are not the same as those of the previous one. The estimator

$$\check{N}_T = g(\acute{N}_1, \acute{N}_2, \acute{N}_3 .... \acute{N}_L) \tag{8.10}$$

is now defined in terms of:

$$\acute{N}_{rh} = 1/n_h \Sigma N_{rhi}$$

where, $N_{rhi} = [y_{rhi} / N_{hp} P_{hi}]$ and P_{hi} denotes the probability associated with the (h, i)-th unit. $\check{N}_{hi}$ denotes the estimator of the same form as equation (8.10) obtained after deleting the (h, i)-th observation from the sample,

$$\check{N}_{(h.)} = \Sigma (\check{N}_{(hi)} / n_h) \; i = 1,........n_h \tag{8.11}$$

$$\check{N}_{(..)} = \Sigma \Sigma (\check{N}_{(h.)} / n) \; h = 1,.......L \tag{8.12}$$

$$\acute{N}_{(..)} = \Sigma \check{N}_{(h.)} / L \tag{8.13}$$

To estimate the variance of N_T, we may use any of the following four formula.

$$V_1 (\check{N}_T) = \Sigma (n_h - 1)/n_h \Sigma (\check{N}_{(hi)} - \check{N}_{(h.)})^2 \tag{8.14}$$

$$V_2 (\check{N}_T) = \Sigma (n_h - 1)/n_h \Sigma (\check{N}_{(hi)} - \check{N}_{(h..)})^2 \tag{8.15}$$

$$V_3 (\check{N}_T) = \Sigma (n_h - 1)/n_h \Sigma (\check{N}_{(hi)} - \acute{N}_{(..)})^2 \tag{8.16}$$

$$V_4 (\check{N}_T) = \Sigma (n_h - 1)/n_h \Sigma (\check{N}_{(hi)} - \acute{N})^2 \tag{8.17}$$

In the following hypothetical example we will use the first formula, that is, equation (8.14) only. Here we assumed that the population is divided into $L = 4$ strata, where N_h describe the size of the h-th stratum.

Now, from the data in Table 8.6 (below), the means of the strata are:

$\check{N}_{(h.)} = \Sigma\ (\check{N}_{(hi)}\ /\ n_h) = 3207.08;\ 3413.61;\ 3164.53;$ and 3265.20 respectively

Also we have,

$\check{N}_{(..)} = \Sigma\ \Sigma\ (\check{N}_{(h.)}\ /\ n) = 3262.60$

and

$\acute{N}_{(..)} = \Sigma\ \check{N}_{(h.)}\ /\ L = 3262.6$

Hence, the variance of $\check{N}_T$ by using the following formula is:

$V_1\ (\check{N}_T) = \Sigma\ (n_h - 1)/n_h\ \Sigma\ (\check{N}_{(hi)} - \check{N}_{(h.)})^2$

$= 30818.55 + 58018.81 + 13456.80 + 42919.79$
$= 145213.95$

Therefore, the standard error of the estimate is: 381.07   (8.21)

Table 8.6. Hypothetical Data to Estimate Variance

Selected CDs	M_i	Selection Prob. Of the CD P_i	Selected ED from the CD	Pop^n Of ED	DSE N^*	MiN*	MiDSE/ $m_i p_i$	Ň(hi)
			Stratum 1 = 2750					
CD_{11} = 272	4	.0989	ED_{13}	64	77	308	3114.25	3238.03
CD_{12} = 204	3	.0741	ED_{21}	80	92	276	3724.69	3034.55
CD_{13} = 136	2	.0494	ED_{32}	68	73	146	2955.46	3290.95
CD_{14} = 138	2	.0501	ED_{42}	70	76	152	3033.93	3264.80
			Stratum 2 = 2759					
CD_{21} = 283	4	.1025	ED_{12}	75	96	384	3746.34	3302.70
CD_{22} = 212	3	.0768	ED_{23}	72	91	273	3554.68	3366.59
CD_{23} = 213	3	.0772	ED_{32}	78	94	282	3652.84	3333.87
CD_{24} = 141	2	.0511	ED_{41}	62	69	138	2700.58	3651.28
			Stratum 3 = 2679					
CD_{31} = 192	3	.0716	ED_{11}	67	77	231	3226.25	3143.95
CD_{32} = 197	3	.0735	ED_{24}	78	82	246	3346.93	3103.73
CD_{33} = 262	4	.0977	ED_{34}	64	69	276	2824.97	3277.71
CD_{34} = 195	3	.0727	ED_{42}	74	79	237	3259.97	3132.71
			Stratum 4 = 2775					
CD_{41} = 344	5	.1239	ED_{11}	76	91	455	3672.31	3129.50
CD_{42} = 279	4	.1005	ED_{23}	72	88	352	3502.48	3186.11
CD_{43} = 138	2	.0497	ED_{33}	62	68	136	2736.41	3441.74
CD_{44} = 141	2	.0508	ED_{44}	72	80	160	3149.60	3303.74

8.5.5 Random Groups Method

When we have stratified samples and we wish to estimate the total variance across all strata, then we have the following procedure. Let the sample n_h be divided into K random groups of size m_h for h=1, 2, 3,......, L and let $N_{(i)}$ denote the estimator of N obtained after removing the i-th group of observations from each stratum. Define the pseudovalues

$$\check{N}_i = k\,\check{N} - (k - 1)\,\check{N}^*_i \tag{8.22}$$

Then the estimator of N and the estimator of its variance are

$$\check{N}_R = \Sigma \, \check{N}_i / k \, ; 1 = 1,2,.....k \qquad (8.23)$$

and

$$V(\check{N}_R) = 1/k(k-1) \, \Sigma \, (\check{N}_i - \check{N}_R)^2 \qquad (8.24)$$

For the application of the above variance formula, we will construct the Random Group (RG) by drawing a simple random sample without replacement (srs wor) of size $m_h = n_h/K$ from the parent sample n_h in the h-th stratum, for h=1, 2,...., L. The second RG is obtained in the same way by selecting from the remaining n_h-m_h units in the h-th stratum. The remaining RG's are formed in like manner. If excess observations remain in any of the strata, i. e., $n_h = Km_h + q_h$, they may be left out of the K random groups or added, one each, to the first q_h RGs.

For this example RGs are formed from Table 8.6.

Table 8.7. Constructing Random Groups

Selected CDs	Estimated population from Stratum	RG Estimate
Random Gruop 1		
CD_{14}	3033.94	
CD_{23}	3652.84	
CD_{32}	3346.93	13183.3
CD_{44}	3149.60	
Random Gruop 2		
CD_{13}	2955.46	
CD_{24}	2700.58	
CD_{31}	3226.25	12384.77
CD_{42}	3502.48	
Random Gruop 3		
CD_{12}	3724.69	
CD_{22}	3554.68	
CD_{33}	2824.97	13776.65
CD_{41}	3672.31	
Random Gruop 4		
CD_{14}	3114.25	
CD_{21}	3746.34	
CD_{34}	3259.97	12856.97
CD_{43}	2736.41	

Now for the estimation of the variance by the Jackknife method we construct the following table (Table 8.8).

Table 8.8. Estimating Variance by Jackknife Method

RG	Estimate from RG	$\check{N}_{(\alpha)}$	$\check{N}_\alpha$	$(\check{N}_\alpha - \check{N}_R)$
01	13183.03	13139.01	12784.65	70892.79
02	12384.77	12606.65	14381.73	1771089.80
03	13776.65	13533.92	11599.92	2105363.20
04	12856.97	12921.45	13437.33	149322.73

Therefore

$$V(\check{N}_J) = 1/k(k-1) \sum (\check{N}_\alpha - \check{N}_R)^2 ; \alpha = 1, \ldots\ldots\ldots k \quad (8.25)$$

$$= 4096668.4/12 = 341389.03\$$$

Or S.E. of $V(\check{N}_J)$ = 584.285

The Jackknife was first introduced by Quenouille (1949) as a method of reducing the bias of an estimator of a serial correlation coefficient. Subsequently it has become a powerful general methodology frequently applied to variance estimation.

In our example we were concerned with estimating the size of the population by the dual system estimation. We therefore, constructed a hypothetical data table accordance with the CVS sample design, and applied two version of the Jackknife to the data.

8.6 Model Error

8.6.1 Sources of Error

In Chapter 2 Section 2.4 we defined different types of assumptions of the dual system estimator. Among the assumptions, failure in any of the three independence assumptions leads to error in the model error $N^* - N_{True}$. To reduce the effect of failure of the causal independence and heterogeneous independence assumptions, we will post-stratify the data based on demographic and geographic variables, a technique originally recommended by Chandra-Sekar and Deming (1949). An estimate of the population in each post-stratum is calculated and then all the estimates are summed to give an estimate of the total population. Unless the homogeneity assumption fails, the estimate lies between the census and the truth (Mulry, 1991). Although failure of autonomy (each individual acts alone as to inclusion in the sample and the census) contributes to bias in N^*-N_{True}, research by Wolter (1986c) and Cowan and Malec (1986) has demonstrated that if the other two independence assumptions hold then the bias due to the failure of autonomy is negligible.

8.6.2 Definition

Let $K = x_{11}x_{22}/x_{12}x_{21}$ be the overall cross product ratio and let $\Gamma = K-1$. We will refer to Γ as the model bias factor that reflects failure of the independence assumptions. If the independence assumption holds then $1 = K = K_i$ for $i = 1,2,...N$. The model error may be expressed as

$$N^* - N_{True} = - \Gamma(x_{12}x_{21}/x_{11}) + \varepsilon$$

with the ε the random component of model error that is negligible in this assumption.

8.6.3 Measurement

One way to measure model bias is to compare the PES estimate of population size with an independent estimate and use the differences between the two sets of estimates to make inferces about the magnitude of Γ. For this we will use Demographic Estimates (DE) to compare with our estimate if such are available from the other source; otherwise, this will play little part in our programme of research.

8.6.4 Example

In this example, $\check{N}^*$ is the estimated value by using DSE and DE is the estimated population from the same ED block as used in the case of DSE.

Table 8.9. Estimating Model Error

Selected EDs	$\check{N}^*$	DE	$d = \check{N}^* - DE$
$ED_{1.3}$	3307	3264	43
$ED_{5.2}$	3474	3436	38
$ED_{6.3}$	3159	3118	41
$ED_{9.4}$	3292	3255	37

$E(d) = 39.75$
$V(d_n) = 5.69$ and S.D. $= 2.38$
$V(d_{n-1}) = 7.58$ and S.D. $= 2.75$

8.7 Measurement Error

We will consider all sources of error other than model and sampling errors as measurement error. We also assume that the relative measurement error, $(\dot{N} - N^*)/\dot{N}$, is approximately equal to a linear combination of three error of components,

$$(\check{N}_P - x_{+1})/\check{N}_P + (\check{N}_{CE} - x_{1+})/\check{N}_{CE} + (\check{N}_{CP} - x_{11})/\check{N}_{CP}$$

Now we will examine only the nonsampling error component of each of the above three components.

8.7.1 Estimate of $(\check{N}_P - x_{+1})$

The error committed during the estimate of $\check{N}_P$ is equal to $(\check{N}_P - x_{+1})$ which is the sum of sampling error $\tilde{n}_p$ and non-sampling error n_p.

$$\tilde{n}_p = (\tilde{N}_P - x_{+1}) \tag{8.26}$$

$$n_p = \check{N}_P - \tilde{N}_P = n_{pf} + n_{pa} + n_{pi} \tag{8.27}$$

where

n_{pf} is the error that occurs when interviewers make up fictitious household members.
N_{pa} is the error that occurs due to the misreporting of Census Day address by outmovers and inmovers and
N_{pi} is the matching error that occurs from incorrect imputation of missing data in the P sample.

8.7.2 Estimate of $(\check{N}_{CE} - x_{1+})$

The error committed during the estimate of $\check{N}_{CE}$, that is to estimate x_{1+} is equal to $(\check{N}_{CE} - x_{1+})$ which is the sum of sampling error $\tilde{n}_{CE}$ and non-sampling error n_{CE}.

$$\tilde{n}_{CE} = \tilde{N}_{CE} - N_{CE} \tag{8.28}$$

$$n_{CE} = \hat{E}_E - \tilde{E}_E = c_o + c_e + c_i \tag{8.29}$$

where

c_o arises during the processing of the E sample when respondents are misclassified as to whether they are correctly or erroneously enumerated in the original enumeration.

c_e is the error due to interviewer and

c_i is the error due to imputation.

8.7.3 Estimate of $(\check{N}_{CP} - x_{11})$

The error committed during the estimate of $\check{N}_{CP}$ that is to estimate x_{11} is equal to $(\check{N}_{CE} - x_{11})$ which is the sum of sampling error $\tilde{n}_{CP}$ and non-sampling error m.

$$\tilde{n}_{CE} = \tilde{N}_{CP} - x_{11} \tag{8.30}$$

$$m = \check{N}_{CP} - \tilde{N}_{CE} = m_m + m_a + m_f + m_i \tag{8.31}$$

where

m_m is the error introduced in the matching process.

m_a is the error introduced by respondents reporting the wrong Census Day address.

m_i is the error in assignment of match status that were caused by error in imputation and

m_f is the error introduced due to fabrication.

8.8 Conclusion

In the census as well as in any survey estimates of population are affected by a number of different sources of error. Some of the errors cancel, others do not. When we said that the survey's estimate of the net underenumeration in the census came to 288,000 individuals, clearly this depends on the net error of the estimate. In this chapter we tried to describe a methodology for organizing and summarizing these sources of error which we called 'Total Error Model'. The philosophy of the model is to try to identify all the sources of error, estimate their magnitudes, and assess their effects on the statistics of interest. Mulry and Spencer (1993) describe the model as dynamic as it can incorporate new or alternative information about the kinds or magnitudes of errors as that information becomes available. We illustrated the methodology with the help of hypothetical data. We described how to use the Jackknife method to estimate sampling variance from the CVS sample.

In our model all the sources of error come together. This type of model is very useful for any researcher to identify the sources of error and to develop methods to measure those errors. Using our models, ONS can easily estimates the errors in the CVS and the census and by comparing the results can reach a conclusion about the level of underenumeration.

Chapter 9

Recommendations

9.1 Cost-benefit Analysis

The census is the key source of information for many activities inside and outside Government. Census statistics help to improve public understanding of nations and their communities. The transfer of funds from central to local government is based in part upon the number of people in each administrative area; the location of services such as schools provided by local Education Authorities are based on census statistics. Of course, many other organizations and individuals in addition to Government agencies, business organizations, universities and other research institutions, nonprofit organizations, the media, students, and individual citizens-- make use of census data, often in conjunction with data of their own. It was estimated that approximately 10,000 people in Britain made direct use of the 1981 census statistics and the decisions arising from some of these uses will ultimately touch the lives of millions (Denham and Rhind, 1983). An improved and more efficient census is therefore always important.

Accurate and timely statistics produced from the census with detailed local information is vital for many uses. However, it is well known that a hundred per cent accurate count is impossible, even if we invest unlimited resources, especially in non-response follow-up, and coverage programmes at the later stage of any census or survey add significantly to the total cost with little coverage improvement. The rise in costs does not produce a better census. It is not always cost-effective to increase efforts for highly accurate counts of detailed area and population groups by means of highly intensive operations seeking to make a physical count of every last person (Edmonstone et al, 1995).

The cost of census activities has increased every year. In the U.K. the census is financed by Government and cost about $135 million in 1991. This cost may be increased much more in coming censuses without any gain in census coverage if we fail to handle the expenditure in a cost-effective way. Attempts must therefore be made to investigate methods of analysis which improve census coverage and reduce differential undercoverage as well as gross errors with minimum cost. The main problems are those of allocating resources, that is, how and where we will invest more resources. A few examples may explain the difficulty of the choices that have to be made. How are we to decide the importance of investment in improving the present method against the demands of starting a new method of estimation? How much should be spent on changing the sampling design or improvements in questionnaire and interviewing technique? Should we spend more resources on non-response follow-up, vacant and absent follow-up sampling? Should we attempt to develop a third source of information such as administrative record lists to facilitate the check of the coverage of census counts? What are the criteria to judge that investment on one issue gives more benefit than another? Cost-benefit analysis, therefore, is always useful in assessing the new method, sampling design, non-response follow-up, vacant and absent follow-up sampling or the use of any third source of information like administrative record lists. Here, by benefit I mean

a) improvement in the coverage of the census counts
b) reduction of differential undercount, by age-sex-race-region and
c) reduction of gross-error of the census counts.

Recommendation 1.1: The ONS should give greater importance to cost-benefit analysis of the overall census and CVS methodology to implement decisions for the future censuses and CVS. If ONS wants to use a Dual-System method in the 2001 census coverage evaluation programme they should develop a plan as to how they will implement the methods and how much it will increase the cost with the increase of coverage.

9.2 Design Issues

The CVS methodology of estimating census coverage does not have the opportunities to estimate people who were missed by both the methods viz, census enumeration and CVS estimates. Moreover, CVS methodology in many cases depends on the census enumeration and one does not have much idea how this correlation between census and CVS effects the CVS estimates.

To avoid the effect of this correlation error and to estimate the people missed by both the methods, census and survey, the U.S. Bureau of Census adopted dual system estimates method, in which an entirely independent sample is drawn and matched with census enumeration and from the matching result estimate the total figure which we described in Chapter 4. Such a design can produce not only people missed by the both methods, but also improved accuracy for the single most important variable: the count of population, particularly by age, sex, race and regions.

Recommendation 2.1: The ONS should start its research programme on dual system estimates so that they can introduce the methods in 2001 CVS. Efforts should include

studies on the stratification, sampling design, different matching criteria and imputation techniques. A test programme before the census may help to take more appropriate steps.

In the CVS six samples were selected for the estimation of undercount, five of which were dependent on the census count. Clearly, this sample design has some bad effect on the estimate. First one needs to wait until the completion of all census work, which may cause memory error. Second, as the samples are dependent, the estimate may be affected by correlation. We believe a better estimate can be achieved by applying a CensusPlus type method. After selecting the ED block, if we re-enumerate all the housing units listed by the CVS interviewers independently instead of drawing five different samples from the census count, this may give a better chance of enumerating missing persons as well as erroneous enumeration.

Recommendation 2.2: For a better CVS estimate of the population an independent sample is essential. One way of doing so is to adopt a CensusPlus type sampling method. This type of sampling method is likely to improve the coverage as well as erroneous enumeration.

In chapter 5 we discussed using log-linear models to estimate the missing people as well as the total population of the country by using information from three different sources, mainly, census enumeration, survey information and administrative record lists. Ericksen and Kadane (1985) suggested developing a megalist from all available sources of information like health service records, driving license record lists, post office addresses lists etc. to use as a third source of information. Leggieri (1994), in his paper presented in the U.S. Census Bureau's conference, suggested the creation and maintenance of a master list of addresses over the decade. He also gives the detailed criteria of developing the master list.

Recommendation 2.3: To cope with the development of new technology to improve estimation of the undercoverage of the census a third source of information which is completely independent of census and survey is necessary. Administrative record lists is one way to have that information. The ONS should take necessary steps for the development of a master address list by combining all the available administrative lists.

The address list developed by the census office generally served only limited purposes after the census operation was completed. The master address lists which contain the information for each residential unit that is necessary to contact households by mail, by telephone or by personal visit can be more useful. The future scope of using this type of address list within the boundary of maintaining confidentiality is also greater and necessary for the census office as well as other Governmental and business organizations.

Recommendation 2.4: The ONS should developed a structure so that the master address lists could be shared among Government and Local Government agencies, research and business organizations for use under appropriate conditions.

9.3 Response and Coverage Issue

The most important finding of the 1991 demographic estimates is that 1.2 million people were missed from the 1991 census count. The reasons for these missing people are not clear. It is known from the CVS that some households were completely missed from the ERB lists, and hence all the people of the households were missed and some other people missed from the listed households either by chance or deliberately. Mobile people especially young men between 20-34, students and people with a second residence increased these numbers. Desire to avoid community charge may be another reason of underenumeration.

In Britain the census is compulsory and failing to comply with the census regulations or the provision of false information, could result in a fine up to 50 pounds. In addition to these, the rule of the strict confidentiality of the collected data convinced most of the population to participate in the census as a national duty. A sizeable and perhaps increasing proportion appears to be motivated primarily by local interests and appeals, demand control over portions of the census process or outcomes or require specialized help or media in order to participate (Steffy et al, 1994). Many people need help with the skills to respond to the census questionnaire. Many other are simply unmotivated or distrustful of Government efforts to collect information. If the trend of the missing people of 1991 census continue, the 2001 census will face an even larger number of missing people. A greatly improved census procedure even may not yield greatly improved outcomes. A recognition of the present problem of counting the population must be incorporated into the planning for 2001 in order to develop viable strategies and the organization and political awareness to implement them.

I am not sure what kind of research programme is now running or under consideration to improve the 2001 census, by the ONS. However, I know that after the 1991 investigation ONS are concerned about the imputation technique, identification of distinct households in shared dwellings, undercoverage of residents in large households, treatment of individuals who were away from home on census night and some definitional problems like household space etc. (Heady et al, 1994).

For all these issues development of various research programs is essential. Research aiming to reduce coverage error, by improving ERB lists, have the potential to produce important innovations and possibly reduce the differential undercount. Research to improve the ED boundary problems will not only reduce the discrepancy between the judgements of census office staff and census field staff about which ED particular buildings should be assigned to, but will also help to create a more accurate list of all buildings belonging to each ED.

Coverage error may be improved further by listing all buildings at least two weeks before the census and then comparing the list with the master address lists as we mentioned in chapter 5 and will be discussed further.

From the CVS quality check sample (census co-operation sample) 155,000 people were found who were not enumerated in the 1991 census, and were in households that returned their questionnaire. This figure is the net outcome of partially cancelling gross enumeration errors (Heady et al, 1994). The coverage error which occurred within households, that is, the error which occurred due to mistakes in the households while completing the questionnaire is known as response error (Steffey et al, 1994). To improve the coverage error as a whole, the reduction of response errors within the households is necessary, because it is household respondents who implement the residence rules as they fill out the census questionnaire or talk with a census enumerator. One way of improving response errors is to improve the questionnaire. If the respondent understands the questions properly maybe s/he will be able to give more accurate answers to the question. Good news is that ONS will carry out a census test on 15 June 1997 in which some of the new questions as well as revised wording of the previous census questions will be tested (Census News No 37, 1996).

Reducing omissions and erroneous enumeration has the general benefit of reducing the variance of coverage. If a method can be developed to improve the quality of the initial count, by improving ERB lists of all buildings, sending advance letters or by reminder cards it will help to control errors as well as increase the credibility of subsequent estimates.

Most confusion arose in the 1991 census from the housing space definition. In some cases census interviewers did not realise that there was more than one household space and failed to give household a questionnaire and in other cases the respondent did not make sure exactly who would be included in the form (Heady et al, 1994).

Several research projects which address, for example, the questions that must be answered in order to increase the accuracy of coverage within households, techniques of checking the census questionnaire during collection time etc. are essential to improve the next census.

Recommendation 3.1: To reduce coverage errors by reducing missing residential housing units ONS should undertake a research programme to improve the ERB listing of all building units within ED blocks.

Recommendation 3.2: ONS should undertake a programme to check and match the ERB listing of residential buildings with a third source especially master address lists created from all available lists of addresses.

Recommendation 3.3: ONS should undertake a programme of research to reduce the coverage errors within the households by reducing response errors. This can be done by improving the census questionnaire.

Recommendation 3.4: ONS should undertake a programme to understand the real Causes of underenumeration especially in those places where the rate of underenumeration is high.

Recommendation 3.5: ONS should undertake a programme to develop good working relations with the local authority so that their resources can be used if necessary especially in hard to enumerate areas.

Recommendation 3.6: ONS should undertake a programme to develop a plan to enumerate people living in the street or rough. Perhaps a better way of doing this is to estimate on a sample basis, which requires an appropriate sampling plan.

Recommendation 3.7: To create census awareness, the ONS should conduct a census media campaign. Besides advertising in the radio and television, working irectly with local and regional agencies, undertaking paid media research, and campaigning door-to-door in hard-to-enumerate and underdeveloped areas may improve the coverage and quality of the census.

9.4 Sampling and Statistical Estimation

Sampling and subsequent estimation always offer some advantages over enumerating an entire population. For example, trying to obtain information from everyone in a large population is usually expensive. The quality of data also reduces with the size of the population. A well conducted sample survey usually provides more accurate information than a survey that attempts to collect data from an entire population (Kish, 1965).

In the 1991 U.K. census, in the case of absent households, census interviewers left census questionnaires at these addresses, along with a leaflet inviting the absent residents to complete the questionnaire and post it back to the census offices when they returned home. Many of the absent households did complete and return the questionnaire to the census office and many did not.

For the absent households which did not complete a census questionnaire, the census office used information relating to similar addresses in the same area to impute the characteristics of the absent household, on the basis of the information collected by the enumerators from the neighbours or from the occupants themselves before they left.

This problem can be dealt with using sampling. Instead of trying to matching absent households with a similar household of the same area, ONS could follow-up only a sample of such housing units (most likely 10 to 30 percent) within a week or two weeks from the census day. Data from housing units sampled for absent household follow-up would allow estimation of counts and characteristics of absent households who were not sampled.

Recommendation 4.1: Sampling for absent households could produce more accurate and timely estimates. The process may also reduce the cost of the census. A test project before the next census will help to tackle the practical problems of adopting the techniques.

Demographic estimates show that in Great Britain approximately 1,300,000 people were missed from the 1991 census enumeration. When adjusted for definitional differences, this figure reduces to 1,200,000. Among these missing people approximately 1,150,000 were under 45 years old. Comparison with demographic estimates also shows that underenumeration was much more marked among young men than among young women (Heady et al, 1994).

From the beginning of the post enumeration survey, all the survey results have demonstrated the existence of an overall undercount. The post enumeration survey also found that there is a differential undercount, i.e., certain groups, such as young men between ages 20-34 and certain areas, such as inner cities, are systematically undercounted relative to other groups and areas in conventional census enumeration. Despite careful investigation of the CVS methods and details of its design ONS failed to explain all the shortfall of the census underenumeration.

From the past two CVS surveys and from other research ONS was able to identify some of the problems that affect the census counts. Using this experience ONS expects to improve enumeration in the next census. However it is unlikely that these experiences will reduce the underenumeration completely, especially when ONS officials themselves are unable to explain many problems of underenumeration. Therefore the need for the CVS estimates in 2001 will be at least as great as in previous censuses.

To reduce the underenumeration as well as differential undercoverage by estimating the magnitude of such differentials, the U.S. Bureau of Census is planning to use a method which they called Integrated Coverage Measurement. This integrated coverage measurement includes the use of samples, statistical estimates based on these samples, and statistical modelling, as an essential part of the census, with the other census-taking operations. For 2001, ONS can think about this type of operation, especially to deal with the absent and vacant households units as a test case.

Recommendation 4.2: To deal with the differential undercount, use of some statistical technique is essential. We propose that ONS undertake a research project like that of Integrated Coverage measurement for the next census.

Demographic analysis as a tool for coverage evaluation is well developed and has been using by many countries around the world to assess the completeness of coverage in censuses. For census evaluation, demographic analysis first develops methods of estimation using the basic demographic identities relating population to births, deaths, immigration and emigration (Robinson et al, 1993). Traditional methods for demographic analysis yield estimates of the national population cross-classified by age-sex-race. The estimated values are then compared with the corresponding census counts, to yield a measure of net undercount.

In the 1991 census ONS decided to use demographic estimates as best estimates of census undercoverage (Population Trends, No. 71, 1993). This is because from the assessment of the possible sources of errors in the demographic estimates it was concluded that there was no reason to think the figures produced by demographic methods were overestimated. On the other hand, it was possible that persons missed by the census were also missed by the CVS and hence CVS underestimated the number missed by the census (Heady et al, 1994). Another reason for accepting the demographic estimates as a best estimate is that, census figures even after correction by CVS, showed an implausibly high ratio of women to men among young adults. This may be due to differential response to the census and CVS (Population Trends, No. 71, 1993).

Further research and necessary steps to use demographic estimates for the 2001 census coverage evaluation purposes will be a cost effective investment that could pay long-term dividends beyond the contributions to census coverage and evaluation. It is essential to use other sources of information with the traditional demographic data to produce reliable estimates of the population cross-classified by age, sex, and race. Methods to estimate subnational geographic areas by demographic analysis also need an appropriate research programme.

Recommendation 4.3: We believe use of demographic estimates is essential to estimate coverage and evaluation of the census. It will be very useful if necessary steps are taken to have demographic estimates within a short time of the census enumeration result. An appropriate research programme is essential to use information from other sources like administrative record lists with traditional demographic data to improve the quality of data like migration data which will in turn improve accuracy and reliability of the estimates. A research programme is also necessary to develop methods of estimation for subnational geographic areas.

Data ranging from simple information of population characteristics to complex tabulations and sample microdata files are produced from census information. For the analytical uses of these data, users of census statistics require guidance on the range of possible errors in the basic counts of population and in the counts of subpopulations within the total (Rhind et al, 1983). Errors arise because some people are missed entirely and some are double counted, and because characteristics such as age and occupation are wrongly recorded, wrongly coded or wrongly key punched. It therefore becomes the responsibility of the data producers to facilitate the estimation of uncertainty.

OPCS used standard errors to estimate the uncertainty of their estimates. But sampling error is not the only error of census counts. Mulry and Spencer (1991, 1993) developed a total error model for estimating the uncertainty in adjusted census based on 1990 census and PES. Their models take account of all possible sources of errors which they divided broadly into three categories, sampling errors, model errors, and measurement errors. Similar models may be useful for evaluating uncertainty for the 2001 U.K. census. After measuring the uncertainty they should be published with the census statistics in a manner that allows users to understand the risk of using the statistics.

Recommendation 4.4: To estimate all sources of errors in censuses and surveys ONS should develop models like that of Total Error Model by Mulry and Spencer (1991, 1993). They should publish summary measures of uncertainty in a standard statistical way such as average coefficient of variation along with census statistics.

9.5 Creating Administrative Record Lists

Present experience from the CVS and demographic estimates and from the extensive research work by Statistics Canada into the potential use of a variety of administrative records for small area estimation make many people believe that there are significant benefits to be obtained from greater use of administrative records both in the decennial census programme and in the programmes that provide current demographic data. However, the effective use of administrative data by the ONS requires a legal right to access. Establishing good working relationship between ONS and the custodians of administrative records, and reasonable assurance of continued access to data that are suitable for the intended statistical uses are also necessary. Moreover some modification of the present record system by addition to or changes in content may be necessary to increase the effective use of the records for statistical analysis.

The Home Office is planning to introduce Identity Cards for all citizen more than 18 years old. Administration of the Identity Card (ID) system will require the creation and continuous updating of enrollment records for all persons who are given an ID with information about their identities, current addresses, characteristics etc..Potentially, these ID records for people 18 and over could provide a more complete coverage of the U.K. population with current information about each person's location and demographic characteristics. If the Home Office and ONS agree to work together these records could be used more effectively for the coverage of population by including some additional information like race and ethnicity. Besides this, ONS can also take necessary action to utilize NHS records, child benefit records, social security records etc. for the improvement of coverage of population.

As we mentioned before, legal access to the data is necessary for their effective use. ONS should put recommendations on how they will maintain the privacy and confidentiality of all these records when used for statistical analysis. Moreover, acceptance of statistical uses of administrative records by those who provide information about themselves to the programme agencies is also vital. Questions naturally arise, do people accept the use of information about themselves for statistical purposes that are not directly related to the purposes for which they supplied their information? Will the confidentiality of their data be adequately protected? All these questions and many more should be answered before planning to use administrative records for census evaluation and coverage improvement.

We believe it is necessary to proceed with a public debate about the uses of administrative records for census evaluation and coverage improvement. Discussions are likely to be more productive if they focus on specific uses of administrative records, such as improvement of coverage in future censuses rather then on broad philosophical questions (Steffy et al, 1994).

We believe it will be very useful if ONS takes the initiative to organize a conference on `Statistical uses of Administrative Records' and invite all the agencies and custodians of administrative records to participate in the discussion of the conference. The U.S. Bureau of the Census in 1993 organized such a conference, the outcome of which was very useful. Many of the custodians of administrative records system recognized that sharing their records for statistical uses would have benefits. Many custodians said legal authorization is essential from an appropriate authority before sharing the records. They were all conscious of the need to inform data subjects about how their data would be used and to inform the public about the benefit and risks associated with data sharing (Bureau of the Census, 1994).

Recommendation 5.1: To achieve the benefit of administrative records at a minimum cost, mutual working relations between ONS and other custodians of records is essential. To take full advantages of the new ID system ONS should participate with the Home Office actively in the development of content and access provisions for these record systems.

Recommendation 5.2: The ONS in co-operation with other agencies and organizations should undertake research programme on public views about statistical uses of administrative records, which are recorded for other reasons. The research should focus on public reaction to very specific administration record use scenarios, rather then on general questions of privacy.

Recommendation 5.3: The ONS should put recommendations to the appropriate authority which on the one hand will give them the legal right for the use of administrative records and on the other hand gives strong protection of the confidentiality of individual information.

9.6 Small Area Estimates of Population

Accurate small-area estimates of various characteristics of the population depend not only on the methods of estimation but also on the availability of accurate demographic and related data for small areas--cities, counties, districts, neighbourhood, and other small geographic areas. As both the public and private agencies have expanded and refined their uses of the data, the demand for accurate, timely and consistent data has steadily increased from counties to neighbourhood. Though the census is the richest source of small area data, it does not cover all kinds of data and also is not available at all times. If any area experiences any economic or rapid population change, in the few years after the census, the decennial census information for that area may no longer provide accurate information about the people, their education, income, and other characteristics. No matter how accurately the census data were collected, they lose accuracy as time passes. Besides this, errors also occur during the time of census enumeration. Data collected in the census especially at ED level have non-response errors, response errors, geocoding errors, item non-response errors and imputation errors.

In Chapter 6 we discussed a model based estimation method for small-area populations. Another popular way of estimating small area population characteristics is survey based data in which estimates are derived directly through the collection of data. Compared to model based estimates, survey based estimates are costly, require more man power, resources, highly trained interviewers and time. Model based estimates, in contrast, generally rely on existing data and hence are less costly. The reliability of the model based estimates largely depends on the accuracy of the existing data. Apart from censuses and surveys these data can also be collected from administrative lists. These administrative lists can supplement census counts to improve and update census data in the small areas. Statistical demographers have developed several competing methods from these administrative lists for population counts of small areas. However, these methods have been essentially accounting procedures specialized for population counts, with the notable addition of the ratio-correlation method, which has wider application (Purcell et al, 1979).

Recommendation 6.1: We believe information gathered from three sources, census, survey and administrative lists will produce improved intercensal estimates for small areas. We therefore suggest that ONS should undertake a research project to develop methods of estimation by using data from the above three sources. They should also work to improve the quality and develop appropriate recording systems for administrative data so that they can be used for the census.

9.7 Hard-to-Enumerate Population

The main objective of the census is to count all the people in the country present on census night. The census office can do this in two ways. First they can use the same method for all geographical areas and demographic groups. Second, they can use different techniques for different geographical areas and demographic groups to achieve the same population coverage in every area and group. Now, the question is which method is better. As the main goal of census is to measure the population of all geographical areas and demographic groups accurately, obtaining equal coverage clearly takes priority over using the same methods in every area.

In the 1991 CVS to tackle the 'Hard-to Enumerate Population' all the EDs of England and Wales were graded according to the expected difficulty of carrying out the enumeration based on the estimated numbers of non-residential properties, multioccupied addresses, person in communal establishment, expected language difficulties and the area size,. An ED which was assumed easy to enumerate was graded 'A' through to grade 'G' the most difficult one. For sample selection an ED of grade 'A' received a weight of 1, an ED of grade 'B' received a weight of 2; 'C and D' a weight of 3; and 'E, F, and G' a weight of 4. So a grade B ED in Inner London estimated to contain 190 households was weighted to 380 households (Heady et al, 1994).

In the 1990 census, the U.S. Census Bureau conducted an alternative enumeration in 29 sample ares of the U.S. and Puerto Rico called the Ethnographic Evaluation Project. These areas had high concentration of particular minority groups and a large number of illegal immigrants who are known to fear participating in the census. The project report (de la Puente, 1993), described five sources of undercount or overcount: 1) irregular and complex living arrangements, 2) irregular housing, 3) residential mobility, 4)distrust of Government, and 5) limited English proficiency.

All of these causes of undercoverage/overcoverage suggest that different methods are needed for improving census coverage. For example, the undercoverage due to English proficiency can be reduced by native speakers or multilingual enumerators who can communicate effectively and have the capacity to induce the respondent to trust the enumerator.

Recommendation 7.1: The ONS should undertake research, first to identify all the causes of Hard-to-Enumerate populations and second to develop strategies to count or estimate the population of each of these problem area.

9.8 Conclusion:

The main challenge facing the U.K census in 2001 is to find out the extent of the undercount and the possible causes of the undercount. The Office for National Statistics has responded to the challenge by undertaking a Census Test which will be held on 15 June 1997, as part of the planning for the census in 2001. The major objective of the test is to develop field procedures to improve the level of coverage and quality of census data. Other objectives are to Assess the public response to new questions, question wording and different styles of forms design(Census News No. 36, 1996).

In this Chapter we discussed some of the issues which we believe need to be decided before the 2001 census for the improved coverage of the 2001 census within the allocated funds. It is fruitless to allocate more funds to continue trying to count every last person with traditional census methods of physical enumeration. Research on how funds will be allocated by assessing the new methods, sampling design, questionnaire development or using sampling estimates is, therefore, always important.

The Office for National Statistics acknowledged that the present methods of validation need to be improved for 2001. The main weaknesses of the present methods are that they are unable to estimate people missed by both census and survey and cannot estimate undercount by race, age, sex and regions. This situation can be improved by assuming that the CVS is independent of the Census which in turn will give the facility to use the Dual System Estimation technique to estimate the undercoverage of the census.

Finally, we believe an improved ERB supplemented by a second source, such as administrative record lists, can achieve significant increase in the coverage of the census while a friendly questionnaire will help to reduce the response error and improve data quality. We also believe that the present plan of reviewing data collection methods and procedure by ONS is a timely and appropriate step to increase the quality of census data.

Appendix 1

One of the important assumptions of the C-D estimator is that the two methods of data collection are independent. That is they assume that the chance of an event being recorded by the census does not influence the chance of being recorded by the sample survey. This implies zero correlation and so the persons missed by both methods are estimated as

$$X_{22} = x_{12} x_{21} / x_{11} \qquad (9.1)$$

Hence the total estimated number of persons, $\check{N}_T$ is

$$\check{N}_T = x_{11} + x_{12} + x_{21} + x_{22} \, [= x_{12} x_{21} / x_{11}] \qquad (9.2)$$

In reality the assumption of independent collection is unacceptable. Greenfield (1975), El-Sayed Nour (1982), and many others argue that, in particular where the source of data is a human population, there are many possible reasons for which data can be missed systematically by both methods of data collection.

The correlation between the two methods is,

$$r_x = (x_{11}x_{22} - x_{12} x_{21})/[(x_{11} + x_{12})(x_{11} + x_{21})(x_{12} + x_{22})(x_{21} + x_{22})]^{1/2} \qquad (9.3)$$

Based on real observations, Chandra-Sekar and Deming (1949), Jabine and Bershad (1968) and Greenfield (1975), were convinced that the association between these two collection systems is positive. Chandra-Sekar and Deming, therefore, suggested that their method will give better result if applied to homogeneous sub-groups of the data and that the total estimate be obtained by building up from these sub-groups. The underlying argument is that if the association for each sub-group is near zero, while the association for all sub-groups combined is not zero, then a less biased estimate of x_{22} will result.

Greenfield's Method

Greenfield (1975) argues that while the method of sub-grouping offers an improved estimate, it still suffers from the defect that independence within sub-groups is assumed. He therefore, proposed that the C-D estimate of the number of missed events should be regarded as a minimum estimate. The assumption of zero correlation upon which the C-D estimate is based can no more be justified in theory as the minimum value of the correlation than could assumptions of values of, say, -0.05 or +0.05. Nevertheless, there is some empirical evidence that it tends to understate the value of the true correlation and hence to understate the number of missed events, which does accord with intuitive expectations. He therefore proposed a procedure for estimating upper and lower limits to the value of the missing cell. His proposed estimator for the lower limit to the value of x_{22} is the same as that of Chandra-Sekar and Deming (1949) which is

$$x_{22} = x_{12} x_{21} / x_{11}$$

An upper limit to the value x_{22} is derived by taking r_x as Pearson's correlation co-efficient and then writing the equation in the quadratic form for x_{22} and solving for x_{22} as

$$x_{22} = -1/2B + (A + B^2/4)^{1/2} \tag{9.4}$$

The value of r_x, of the above equation is replaced by $\check{r}_x$ which is the mean of algebraic maximum and minimum value of r_x. The technical details are given in Appendix 1.1.

Appendix 1.1

An upper limit to the value of x_{22} is estimated by writing the equation (9.3) in the quadratic form for x_{22} (Appendix 1.2) and solving for x_{22} as

$$x_{22} = -1/2B + (A + B^2/4)^{1/2} \tag{9.5}$$

using the positive root where

1. $A = (x_{12}x_{21})[x_{12}x_2 - r_x^2(x_{11} + x_{12})(x_{11} + x_{21})]/ r_x^2(x_{11} + x_{12})(x_{11} + x_{21}) - x_{11}^2$

$$\tag{9.6}$$

2. $B = [r_x^2(x_{12} + x_{21})(x_{11} + x_{12})(x_{11} + x_{21}) + 2x_{11}x_{12}x_{21}]/r_x^2(x_{11} + x_{12})(x_{11} + x_{21}) - x_{11}^2 \tag{9.7}$

3. $x_{12} \geq 0 \,; x_{21} \geq 0 \,; x_{22} \geq 0 \,; x_{11} \geq 0$

Again from equation (9.3) Greenfield estimates the algebraic maximum value of r_x, $r_{x(max)}$ (Appendix 1.3), is given by the non-negative value of

$$r_{x(e)(max)} = [x_{11}Z - x_{12}x_{21}]/[(x_{11} + x_{12})(x_{11} + x_{21}) \, (x_{12} + Z)(x_{21} + Z)]^{1/2} \tag{9.8}$$

which is the empirical maximum value of r_x, applicable where the variable can only assume the values one or zero for a given member of the set at a given time. Where N_T is the total population or sample size,

$$Z = N_T - x_{11} - x_{12} - x_{21} \tag{9.9}$$

and therefore Z constitutes the maximum possible value of x_{22} (here, Z is the number of members for which no event was recorded by both the methods). The subscript (e) denotes 'empirical' and will be used throughout the following whenever equation (9.8) has been employed in arriving at the particular estimate concerned.

In the case that the variable is not constrained to the values one or zero, then the theoretical maximum value of r_x must be employed in equation (9.12). This is the non-negative value of

$$r_{x(t)(max)} = x_{11} / [(x_{11} + x_{12})(x_{11} + x_{21})]^{1/2} \tag{9.10}$$

where the subscript (t) denotes ``theoretical'' and will be used whenever equation (9.10) has been employed in arriving the particular estimate concerned. By definition,

$$r_{x(e)(max)} \leq r_{x(t)(max)}$$

and since x_{22} is a monotonically increasing function of r_x, the estimated value of x_{22} using the empirical maximum of r_x will be less than or equal to that using the theoretical maximum, the magnitude of the difference depending on the difference $r_{x(e)(max)}$ and $r_{x(t)(max)}$, which in turn on the ratios of x_{12} and x_{21} to Z and to x_{11}.

The minimum possible value of r_x, $r_{x(min)}$, is given by the non-positive value of

$$r_{x(min)} = -[x_{12}x_{21} / [(x_{11} + x_{12})(x_{11} + x_{21})]]^{1/2} \tag{9.11}$$

From the above two values of $r_{x(max)}$ and $r_{x(min)}$, $\check{r}_x$, is estimated as follows:

$$\check{r}_x = 1/2 [r_{x(max)} + r_{x(min)}] \tag{9.12}$$

This estimated value $\check{r}_x$ is then substituted in equation (14) and (15) and we get the final equation of x_{22} which is:

$$x_{22} = -1/2B + (A + B^2/4)^{1/2} \tag{9.13}$$

using the positive root where

1. $A = (x_{12}x_{21})[x_{12}x_2 - r_x^2(x_{11} + x_{12})(x_{11} + x_{21})]/ r_x^2(x_{11} + x_{12})(x_{11} + x_{21}) - x_{11}^2$

$$(9.14)$$

2. $B = [r_x^2(x_{12} + x_{21})(x_{11} + x_{12})(x_{11} + x_{21}) + 2x_{11}x_{12}x_{21}]/r_x^2(x_{11} + x_{12})(x_{11} + x_{21}) - x_{11}^2$

$$(9.15)$$

Appendix 1.2

The derivation of equation (9.4)

$r_x^2 = (x_{11}x_{22} - x_{12}x_{21})^2/[(x_{11} + x_{12})(x_{11} + x_{21})(x_{12} + x_{22})(x_{21} + x_{22})]$

or, $r_x^2(x_{12} + x_{22})(x_{21} + x_{22}) = [x_{11}^2x_{22}^2 - 2x_{11}x_{22}x_{12}x_{21} + x_{12}^2x_{21}^2]/[(x_{11} + x_{12})(x_{11} + x_{21})]$

or, $r_x^2(x_{12} + x_{22})(x_{21} + x_{22}) - [(x_{11}^2x_{22}^2 - 2x_{11}x_{22}x_{12}x_{21})/[(x_{11} + x_{12})(x_{11} + x_{21})]$
$= (x_{12}^2x_{21}^2)/[(x_{11} + x_{12})(x_{11} + x_{21})]$

Or, $(x_{12}^2x_{21}^2) = (r_x^2x_{12}x_{21} + r_x^2x_{12}x_{22} + r_x^2x_{21}x_{22} + r_x^2x_{22})(x_{11} + x_{12})(x_{11} + x_{21})$
$- x_{11}^2x_{22}^2 + 2x_{11}x_{22}x_{12}x_{21}$

Or, $x_{22}^2[P] + x_{22}[Q] = x_{12}^2x_{21}^{21} - r_x^2x_{12}x_{21}[R]$

where

$P = [r_x^2(x_{11} + x_{12})(x_{11} + x_{21}) - x_{11}^2]$

$Q = [(r_x^2(x_{12} + x_{21})(x_{11} + x_{12})(x_{11} + x_{21}) + 2x_{11}x_{12}x_{21}]$

$R = [(x_{11} + x_{12})(x_{11} + x_{21})]$

Therefore

$x_{22}^2 + x_{22}[r_x^2(x_{12} + x_{21})(x_{11} + x_{12})(x_{11} + x_{21}) + 2x_{11}x_{12}x_{21}]/[r_x^2(x_{11} + x_{12})(x_{11} + x_{21}) - x_{11}^2]$

$- [x_{12}^2x_{21}^2 - r_x^2x_{12}x_{21}(x_{11} + x_{12})(x_{11} + x_{21})]/ [(r_x^2(x_{11} + x_{12})(x_{11} + x_{21}) - x_{11}^2]$

$= 0$

Let the coefficient of x_{22} B, and the third factor be A. It then follows that

$x_{22} = -1/2D \pm (C + D^2/4)^{1/2}$

Appendix 1.3

i) The calculation of an algebraic maximum value of r_x

$$r_x = (x_{11}x_{22} - x_{12}x_{21})/[(x_{11} + x_{12})(x_{11} + x_{21})(x_{12} + x_{22})(x_{21} + x_{22})]^{1/2}$$

or, $r_x[(x_{11} + x_{12})(x_{11} + x_{21})]^{1/2}[x_{11}] = (x_{11}x_{22} - x_{12}x_{21})/[(x_{12} + x_{22})(x_{21} + x_{22})]^{1/2}x_{11}$

≤ 1

Therefore

$$r_x \leq x_{11}/[(x_{11} + x_{12})(x_{11} + x_{21})]^{1/2}$$

ii) The calculation of algebraic minimum value of r_x

As r_x is a monotonically increasing function of x_{22}, it follows that its minimum value is when $x_{22} = 0$, which is

$$- (x_{12}x_{21})/[(x_{11} + x_{12})(x_{11} + x_{21})]^{1/2}$$

Appendix 2

El-Sayed Nour Method

El-Sayed Nour (1982), also did not agreed with the suggestion of dividing the data into homogeneous sub-groups and argued that, a better way to dealing with the association bias in estimating x_{22} is to make assumptions concerning the value of the association index r where

$$r = A(x_{11}x_{22} - x_{12}x_{21}) \tag{9.16}$$

where A is an appropriate positive constant. He presents an alternative approach to the estimation of x_{22} which preserves the main characteristics of the C-D technique, but takes into account the lack of independence between the results of the two collection procedures.

He defined the properties of the C-D technique in the context of demographic application and by taking those properties into account, he derived the upper and lower limit of x_{22} and suggested the lower limit as a estimator for x_{22}. His given estimator is

$$x_{22} = [2x_{11}x_{12}x_{21}]/[x_{11}^2 + x_{12}x_{21}] \tag{9.17}$$

The technical details of the estimation procedure are given in Appendix 2.1.

Appendix 2.1

El-Sayed Nour defined the following properties of the dual collection system in the context of demographic application.

a) The two data sources are positively correlated. This means that

$$x_{11}x_{22} - x_{12}x_{21} > 0 \tag{9.18}$$

In addition the question of positive correlation could only arise when

$$x_{21} > 0 \; ; \; (x_{11} + x_{12}) > 0 \; ; \; \text{and} \; (x_{11} + x_{21}) > 0 \tag{9.19}$$

b) The probability that a single event selected randomly from the population will be recorded by a given collection procedure is larger than 0.5. This means that

$$[(x_{11} + x_{12})/(x_{11} + x_{12} + x_{21} + x_{22}))] > 0.5$$

and   (9.20)

$$[(x_{11} + x_{21})/(x_{11} + x_{12} + x_{21} + x_{22}))] > 0.5$$

from which it follows that

$$(x_{11} - x_{22}) > 0$$ (9.21)

c) For the above properties to be consistent, the relationships

i) $x_{12}x_{21} \leq x_{11}^2$
and
ii) $x_{11} = x_{22} = (x_{12}x_{21})^{1/2}$, whenever $x_{12}x_{21} = x_{11}^2$ must hold.

To use the above properties for estimating x_{22} he consider the following inequality,

$$[(x_{11} - x_{22})(x_{11}x_{22} - x_{12}x_{21})]/[x_{22}(x_{11} + x_{12})(x_{11} + x_{21})] > K$$ (9.22)

where K is an appropriate quantity. Whenever the properties cited above for demographic surveys are jointly satisfied, $K \geq 0$.

By writing the equation (9.22) in the quadratic form for x_{22} he found the solution for x_{22} which is equivalent to

$$(x_{22} - X_0)(x_{22} - X_1) < 0$$

or, $X_0 < x_{22} < X_1$,   (9.23)

where both X_0 and X_1 are functions of the unknown K such that

$$X_0 X_1 = x_{12}x_{21}$$ (9.24)

and

$$X_0 + X_1 = [(M^2 + x_{12}x_{21)} - K(M + x_{12})(M + x_{21})]/ M$$ (9.25)

From the equation (9.24) and (9.25}) on the logic that the geometric mean of any two numbers is always smaller than or equal to their arithmetic mean he calculate the value of K which is

$$K \leq [x_{11} - (x_{12}x_{21})^{1/2}]^2 / [(x_{11} + x_{12})(x_{11} + x_{21})] \tag{9.26}$$

Finding the algebraic maximum for K corresponding to $X_0 = X_1 = (x_{12}x_{21})^{1/2}$ and minimum values of X_0 and X_1 are $(x_{12}x_{21})/M$ and M respectively he gives the following inequalities as the root of X_0 and X_1 is the monotonically increasing and decreasing function of K respectively.

$$[(x_{12}x_{21})]/M \leq X_0 \leq (x_{12}x_{21})^{1/2} \tag{9.27}$$

and

$$(x_{12}x_{21})^{1/2} \leq X_1 \leq M \tag{9.28}$$

For the given pair (X_0, X_1), he write the statement (9.23) i.e. $X_0 < x_{22} < X_1$ in the following way

$$x_{22} = WX_1 + (1 - W)X_0 \tag{9.29}$$

where

$$W = (x_{22} - X_0)/(X_1 - X_0), 0 < W < 1. \tag{9.30}$$

He then find the upper and lower limit for x_{22} from the derivative of W with respect to X_0 which is

$$(x_{12}x_{21})/ x_{11} (x_{12}x_{21})/ x_{22} - ((x_{12}^2 x_{21}^2)/x_{22}^2 - (x_{12}x_{21}))^{1/2} \tag{9.31}$$

from which

$$[2x_{11}x_{12}x_{21}]/[x_{11}^2 + x_{12}x_{21}] \leq x_{22} \leq (x_{12}x_{21}))^{1/2} \tag{9.32}$$

This equation does not hold in all the situation. To guard against the situation where this equation might not hold, he proposed the lower bound of equation (9.32) as an estimator for x_{22} which is,

$$x_{22} = [2x_{11}x_{12}x_{21}]/[x_{11}^2 + x_{12}x_{21}] \tag{9.33}$$

Appendix 3

Independent variables used in LA regression analysis

1. pmapho=Proportion of male persons present in households on the census night =(L010003/L010001)
2. ps2034=Proportion of persons single in the age group 20-34 = ((L020080+L020091+L020102+L020085+L020096+L020107) /L020001)
3. presbo=Proportion of Black (Black C'bean+Black African+Black other) and other (Indian+P'stani+B'deshi+Chinese) residents = (L060003 +L060004+L060005+L060006+L060007+L060008 + L060009)/ (L060001)
4. prpbnca=Proportion of persons born in Newcommonwealth and Africa =((L070055 +L070058) /L070001)
5. pea2034=Proportion of persons economically active in the age group 20-34 =((L080025 +L080026+L080027+L080028)/L080001)
6. peabo=Proportion of black and other persons economically active = ((L090003 +L090004+L090005+L090006+L090007+L090008 + L090009) /L090001)
7. pmh2024=Proportion of male persons present in the households in the age group 20-24 on the census night =(L110080/L110001)
8. pir1644=Proportion of residents in households with limiting long-term illness in the age group 16-44 =((L120010 +L120013+ L120016) /L120001)
9. pie1644=Proportion of economically active residents in households with limiting long-term illness in the age group 16-44 =((L140010 + L140011+L140012 +L140013)/L140001)
10. pmd2529=Proportion of male migrants (age group 25-29) between districts but within county =((L150179+L150198)/L150001)
11. pmb2529=Proportion of female migrants (age group 25-29) between districts but within county =((L150180+L150199)/L150001)
12. pmogb=Proportion of migrants from outside GB = ((L150014 +L150015) L150001)
13. pmiwah=Proportion of male imputed in the wholly absent households =(L180002/ L180001)
14. poiwah=Proportion of other ethnic (all but white) groups imputed in the holly absent households =((L180047+L180048)/L180001)
15. phtcar=Proportion of households having three or more cars = (L210006/L210002)

16. phhfr=Proportion of households having five rooms =(L220005) /(L220001)

17. prodep=Proportion of dependent persons in the households with residents =(L280361/L280331)

18. pmm2034=Proportion of married male in the age group 20-34 =((L350081+L350092+L350103)/L350002)

19. pem2034=Proportion of married female in the age group 20-34 =((L350086+L350097+L350108)/L350007)

20. pfd3039=Proportion of devorced female in the age group 30-39 =((L350110+L350121)/L350007)

21. prp2024=Proportion of persons in the age group 20-24 = (L390019) / (L390001)

22. prp2529=Proportion of persons in the age group 20-24 = (L290029) / (L390001)

23. prm2024=Proportion of male in the age group 20-24 = (L390020) / (L390002)

24. prf2024=Proportion of female in the age group 20-24 = (L390024)/ (L390006)

25. prolp04=Proportion of lone parents with child(ren) aged 0-4 only = (L400016 / L400001)

26. proprla=Proportion of persons with dependent child(ren)rented houses from local authority or new town =(L460010/L460012)

27. plm6574=Proportion of households with male lone pensioners in the age group 65-74 =(L470015/L470001)

28. plf6074=Proportion of households with female lone parents in the age group 60-74 =(L470057/l470001)

29. pronwot=Proportion of non-white (black all+Indian+P'stani + B'deshi+ hinese) in own (outright) households =((L490042 + L490043+L490044 +L490045+L490046+ L490047+L490048) /L490001)

30. proncot=Proportion of New Commonwealth households with residents =(L490012 / L490001)

31. probiuk=Proportion of birth (inside UK) of households head of the New commonwealth residents in households =(L500146/L500145)

32. prohser=Proportion of households used as a second residence = (L540023/L540001)

33. prp1619=Proportion of persons in the age group 16 to 19 = (L390010 /L390001)

34. prp3044=Proportion of persons in the age group 30-34 = (L390037) / (L390001)

35. prp4559=Proportion of persons in the age group 45-59 = (L390046) / (L390001)

Appendix 4

Table 7.1.3 Three persons live in each house

S. No.	No. Of person live		Male/Female	Age-group	Race
001	03	a	Male	42	White
		b	Female	36	White
		c	Female	05	White
002	03	a	Male	62	Black
		b	Female	59	Black
		c	Male	24	Black
003	03	a	Male	32	White
		b	Male	30	White
		c	Male	31	White
.	.	.	.	.	.
.	.	.	.	.	.
150	03	a	Male	48	White
		b	Female	08	White
		c	Female	04	White

Table 7.1.4 Four persons live in each house

. No.	No. Of person live			Male/Female	Age-group	Race
001	04		a	Male	42	White
			b	Female	36	White
			c	Female	05	White
			d	Male	02	White
002	04		a	Male	62	Black
			b	Female	59	Black
			c	Male	24	Black
			d	Female	12	
003	04		a	Male	32	White
			b	Male	30	White
			c	Male	31	White
			d	Female	29	Blake
.	.		.	.	.	.
.	.		.	.	.	.
150	04		a	Male	48	White
			b	Female	08	White
			c	Female	04	White
			d	Female	46	White

Table 7.1.5 Five persons live in each house

. No.	No. Of person live			Male/Female	Age-group	Race
001	05		a	Male	42	White
			b	Female	36	White
			c	Female	05	White
			d	Male	04	White
			e	Male	02	White
002	05		a	Male	62	Black
			b	Female	59	Black
			c	Male	24	Black
			d	Female	12	Blake
			e	Female	09	Blake
003	05		a	Male	32	White
			b	Male	30	White
			c	Male	31	White
			d	Female	29	Blake
			e	Female	29	White
.	.		.	.	.	.
.	.		.	.	.	.
150	05		a	Male	48	White
			b	Female	46	White
			c	Female	08	White
			d	Female	04	White
			e	Female	04	White

Table 7.1.6 Six persons live in each house

. No.	No. Of person live			Male/Female	Age-group	Race
001	06		a	Male	42	White
			b	Female	36	White
			c	Female	05	White
			d	Male	04	White
			e	Male	05	White
			f	Female	02	White
002	06		a	Male	62	Black
			b	Female	59	Black
			c	Male	24	Black
			d	Female	12	Blake
			e	Female	09	Blake
			f	Female	08	Blake
003	06		a	Male	32	White
			b	Male	30	White
			c	Male	31	White
			d	Female	29	Blake
			e	Female	29	White
			f	Male	02	White
.	.		.	.	.	.
.	.		.	.	.	.
150	06		a	Male	48	White
			b	Female	46	White
			c	Female	08	White
			d	Female	04	White
			e	Female	04	White
			f	Male	01	White

Bibliography

1. Abernathy, J. R., and Lunde, A.S. (1970), "Dual Record Systems: A review of population surveys in developing countries", Background paper for first POPLAB Planning Conference, April 1070, University of North Carolina, Chapel Hill.

2. Afifi, A. A. and Elashoff, R. M. (1966), "Missing Observations in Multivariate Statistics I: Review of the Literature," Journal of the American Statistical Association, **61**, 595-604.

3. Alho, J. M. (1991), "Variance Estimation in Dual Registration Under Population Heterogeneity," Survey Methodology, **17**, 123-130.

4. Alho, J. M. (1990), "Logistic Regression in Capture-Recapture Models," Biometrics, **46**, 623-635.

5. Alho, J. M., Mulry, M. H., Wurdeman, K., and Kim, J. (1993), "Estimating Heterogeneity in the Probabilities of Enumeration for Dual-System Estimation," Journal of the American Statistical Association, **88**, 1130-1136.

6. Anderson, H. (1978),"On Nonresponse Bias and Response Probabilities", Scandinavian Journal of Statistics, **6**, 107-112.

7. Anolik, I. (1989), "The 1987 Post Enumeration Survey," Proceedings of the Survey Research Methods Section, American Statistical Association, 710-715.

8. Bailar, Barbara A. (1985), "Comments on Estimating the Population in a Census Year by E.P.Ericksen and J.B. Kadane," Journal of the American Statistical Association **80**, 109-114.

9. Bailar, B. A., and Bailar, J. C. (1983), "Comparison of the Biases of the 'hot deck' Imputation Procedure with an 'equal weights' Imputation Procedure," in Incomplete Data in Sample surveys, Vol. III: Symposium on Incomplete Data, Proceedings (W.G. Madow and I. Olkin, Eds.). New York: Academic Press.

10. Bailar III, J. C., and Bailar, B. A. (1978), "Comparison of Two Procedures for Imputing Missing Survey Values," Imputation and Editing of Faulty or Missing Survey Data, U.S. Department of Commerce, 65-75. Proceedings of the Survey Research Methods Section, American Statistical Association, 462-467.

11. Bailar, B. A., Bailey, L., and Corby, C. (1978), "A Comparison of Some Adjustment and Weighting Procedures for Survey Data," American Statistical Association, Proceedings of the Survey Research Methods Section, 175-200.

12. Bailey, L. Jansto, A. and Smith, C. (1990), "Assessing the Effects of Imputed Data on Selected Results," American Statistical Association, Proceedings of the Survey Research Methods Section, 249-253.

13. Bailey, L. (1983), "Compensation for Unit Nonresponse in Recurring Surveys," American Statistical Association, Proceedings of the Survey Research Section, 289-294.

14. Banks, M. J. (1977), "An Indication of the Effects of Noninterview Adjustment and Post-Stratification on Estimates from a Sample Survey," Proceedings of the Social Statistics Section, American Statistical Association, 291-295.

15. Bartholomew, D. J. (1961), "A Method of Allowing for `not-at-home' Bias in Sample Surveys," Applied Statistics, **10**, 52-59.

16. Bateman, D. V., Clark, J., Mulry, M., and Thompson, J. (1991), "1990 Post-Enumeration Survey Evaluation Results," Proceedings of the Social Statistics Section, American Statistical Association, 21-30.

17. Beale, E. M. and Little, R. J. (1975), "Missing Values in Multivariate Analysis," Journal of the Royal Statistical Society, B, **37**, 129-145.

18. Bell, W. (1993), "Using Information from Demographic Analysis in Post-Enumeration Survey Estimation," Journal of the American Statistical Association, 1106-1118.

19. Berk, K. N. (1977), "Tolerance and Condition in Regression Computation," Journal of the American Association, **77**, 863-866.

20. Biemer, P. P. (1980), "A Survey Error Model which Includes Edit and Imputation Error," Proceedings of the Survey Research Methods Section, American Statistical Association, 610-615.

21. Birnbaum, Z. W. and Sirken, M. G. (1950), "Bias due to Non-availability in Sample Surveys," Journal of the American Statistical Association, **45**, 98-111.

22. Bishop, Y. M. M., and Fienberg, S. E. (1969), "Incomplete Two-Dimensional Contingency Tables," Biometrics, **25**, 119-128.

23. Bishop, Y. M. M., Fienberg, S. E., and Holland, P. W. (1975), Discrete Multivariate analysis: Theory and Practice, Cambridge, Massachusetts: MIT Press.

24. Bishop, Y. M. M. (1980), "Imputation, Revision, and Seasonal Adjustment," Proceedings of the Survey Research Methods Section, American Statistical Association, 567-570.

25. Blight, B. J. N. (1970), "Estimation from a Censored Sample for the Exponential Family," Biometrika, **57**, 389-395.

26. Blumenthal, S., and Marcus, R. (1975), "Estimating Population Size with exponential Failure," Journal of the American Statistical association, **70**, 913-922.

27. Bogue, D.J. (1950), "A Technique for Making Extensive Population Estimates," Journal of the American Statistical Association,**45**, 149--163.

28. Bogue, D.J. and Duncan, B.D. (1959), A Composite Method for Estimating Postcensal Population of Small Areas by Age, Sex and Colour. National Office of Vital Statistics, Vital Statistics--Special Reports **47**, No. 6.

29. Bogue, D.J., Misra, B.D. and Dandekar, D.P. (1964), "A New Estimate of the Negro Population and Negro Vital Rates in the United States, 1930-1960," Demography, Volume I.

30. Brackstone, G. J. and Rao, J. N. K. (1981), "An Investigation of Ranking Ratio Estimators," Sankhay C., **41**, 97-114.

31. Brass, W. (1971), "A Critique of Method of Estimating Population Growth in Countries with Limited Data," Bulletin, International Statistical Institute, **XLIV**, Book 1, 397-412.

32. Brass, W. (1975), "Methods for Estimating Fertility and Mortality from Limited and Defective Data," Chapel Hill: Carolina Population Center and Laboratories for Population Studies.

33. Breiman, L. (1981), "Issues in Regression Analysis of Incomplete Data," Proceedings of the Statistical Computing Section, American Statistical Association.

34. Britton, M. and Birch, F. (1985), 1981 Post Enumeration Survey, Social Survey Division, OPCS.

35. Buck, S.F. (1960), "A Method of Estimation of Missing Values in Multivariate Data Suitable for Use with an Electronic Computer," Journal of the Royal Statistical Society, Series B, **22**: 302-306.

36. Carver, J. S. (1976), Bibliography on the Dual Record System. An Occasional Publication. International Program of Laboratories for Population Statistics, University of North Carolina at Chapel Hill.

37. Cassel, C. M., Sarndal, C. E., and Wretman, J. H. (1983), "Some Uses of Statistical Models in Connection with the Nonresponse Problem," In Incomplete Data in Sample Surveys, Vol. III: Symposium on Incomplete Data, Proceedings (W. G. Madow and I. Olkin, Eds.) New York: Academic Press.

38. Causey, B. (1987), "Census Adjustment Based on an Uncertain Population Total," SRD Research Report Number: CENSUS/SRD/RR-87/05, Statistical Research Division, Bureau of the Census.

39. Causey, B. (1986), "A Study of Whether Census Adjustment is Worthwhile," SRD Research Report Number: CENSUS/SRD/RR-86/17, Statistical Research Division, Bureau of the Census

40. Chakraborty, P. N. (1963), "On a Method of Estimating Birth and Death rates from Several Agencies," Calcutta Statistical Association, Bulletin **12**, 106-112.

41. Chandra-Sekar, C. and Deming, W.E. (1949), "On Method of Estimating Birth and Death Rates and Extent of Registration," Journal of the American Statistical Association, 44, 101-105.

42. Chapman, D. W. (1976), "A Survey of Nonresponse Imputation Procedures," Proceedings of the Social Statistics Section, American Statistical Association, 1976(1), 145-251.

43. Chapman, D. W. (1982), "Substitution for Missing Units 1," American Statistical Association, Proceedings of the Survey Research Methods Section, 76-84.

44. Chen, T., and Fienberg, S. E. (1974), "Two-dimensional Tables with Both Completely and Partially Cross-classified Data," Biometrics **30**:629-642.

45. Childers, D. R. and Hogan, H. (1990), "Results Of the 1988 Dress Rehearsal Post-Enumeration Survey," Proceedings of the Survey Research Methods Section, American Statistical Association, 547-552.

46. Childers, D., Diffendal, G., Hogan, H., Schenker, N. And Wolter, K.M. (1987), "The Technical Feasibility of Correcting the 1990 Census," Proceeding of the American Statistical Association, Social Statistical Section, 36-45.

47. Citro, C.F. and Cohen, M.L. (1985), The Bicentennial Census: New Directions for Methodology in 1990, Washington, DC: National Academy Press.

48. Clausen, J. A., and Ford, R. N. (1947), "Controlling Bias in Mail Questionnaires," Journal of the American Statistical Association, **42**, 497-511.

49. Coale, A. J. (1961), "The Design of an Experimental Procedure for Obtaining Accurate Vital Statistics," International Population Conference, New York, 372-375.

50. Coale, A. J. (1955), "The population of the United State in 1950 classified by age, sex, and color-a revision of census figures," Journal of the American Statistical Association, 16-54

51. Coale, A. J., and Rives, N. W. (1973), "A Statistical Reconstruction of the Black Population of the United States, 1880 - 1970: Estimates of True Numbers by Age and Sex, Birth Rates and Total Fertility," Population Index, 39, 3-36.

52. Cochran, W. G. (1968), "The Effectiveness of Adjustment by Subclassification in Removing Bias in Observational Studies," Biometrics, **24**, 295-313.

53. Cochran, W. G., and Rubin, D. B. (1973), "Controlling Bias in Observational Studies," A review, Sankhya, **A35**, 417-446.

54. Cochran, W. G. (1977). Sampling Techniques, 3rd ed. New York: Wiley.

55. Cohen, S. B. (1978). A Modified Approach to Small Area Estimation. Unpublished Ph.D. Thesis, University of North Carolina, Chapel Hill, North Carolina.

56. Cohen, S. B. and Kalsbeek, W. D. (1977), "An Alternative Strategy for Estimating the Parameters of Local Areas," 1977 Proceeding of the Social Statistics Section, American Statistical Association, 781--785.

57. Colledge, M. J., Johnson, J. H., Pare, R., and Sande, I. G. (1978), "Large Scale Imputation of Survey Data," American Statistical Association, Proceedings of the Survey Research Methods Section, 431-436.

58. Cormack, R. M. (1993), "The Flexibility of GLIM Analyses of Multiple Recapture or Resighting Data," In Marked Individuals in the Study of Bird Population, J. -D. Lebreton and P. M. North(eds) Basel: Birkhauser-Verlag.

59. Cormack, R. M. (1992), "Interval Estimation for Mark-Recapture Studies of Closed Populations," Biometrics, **48**,567-576.

60. Cormack, R. M. (1989), "Log-Linear Models for Capture-Recapture," Biometrics, **45**, 395-413.

61. Cormack, R. M. (1981), "Loglinear Models for Capture-Recapture Experiments on Open Populations," In The Mathematical Theory of the Dynamics of Biological Populations II, R. W. Hiorn and D. Cooke (eds), London: Academic Press.

62. Cormack, R. M., and Jupp, P. E. (1991), "Inference for Poisson and Multinomial Models for Capture-Recapture Experiments," Biometrika, **78**, 911-916.

63. Cowan, Charles D. and Bettin, Paul J. (1982), "Estimates and Missing Data Problems in the Postenumeration Program," technical reports, U.S. Bureau of the Census, Washington DC.

64. Cowan, Charles D. and Fay, Robert E. (1984), "Estimates of Undercount in the 1980 Census," Proceeding of the Survey Research Methods, Section of the American Statistical Association, pp. 560-565.

65. Cowan, C.D. and Malec, D. (1986), "Capture-Recapture Models When Both Sources Have Clustered Observations," Journal of the American Statistical Association, **81**, 347-353.

66. Cox, B. G., and Folsom, R. E. (1978), "An Empirical Investigation of Alternate Item Nonresponse Adjustments," Proceedings of the Survey Research Methods Section, American Statistical Association, 219-223.

67. Cressie, N. (1989), "Empirical Bayes Estimation of Undercount in the Decennial Census," Journal of the American Statistical Association, **84**, 1033-1044.

68. Cressie, Noel (1987), "Empirical Bayes Estimation of Undercount in the Decennial Census," Preprint 86-58, Iowa State University, Department of Statistics.

69. Crosetti, A.H. and Schmith, R.C.(1956), "A Method of Estimating the Intercensal Population of Countries," Journal of the American statistical Association, **51**, 587--90.

70. Dagenais, M. G. (1971), "Further Suggestions Concerning the Utilization of Incomplete Observations," Journal of the American Statistical Association, **66**, 93-98.

71. Daniel, W. W., Schott, B., Atkin, F. C. and Davis, A.(1982), "An Adjustment for Nonresponse in Sample Surveys," Educational and Psychological Measurement, **42**, 57-67.

72. Darroch, J. N. (1958), "The Multicapture Census, I. Estimation of a closed Population," Biometrika, **45**, 343-359.

73. Darroch, J. N., Fienberg, S. E., Glone, G. F. V., and Junker, B. W. (1993), A Three-Sample Multiple-Recapture Approach to Census Population Estimation with Heterogeneous Catchability," Journal of the American Statistical Association, **88**, 1137-1148.

74. Das, G. (1964), "On the Estimation of the Total Number of Events and the Probabilities of Detecting an Event from Information Supplied by Several Agencies," Calcutta Statistical Association, Bulletin **13**,89-100.

75. David, M. H., Little, R. J. A., Samuhel, M. E., and Triest, R.K. (1983), "Imputation Methods based on Propensity to Respond," American Statistical Association, Proceedings of the Business and Economics Section.

76. David, M. H., Little, R. J. A., Samuhel, M. E., and Triest, R. K. (1986), "Alternative Methods for CPS Income Imputation," Journal of the American Statistical Association, 81, 29-41.

77. Davis, M., Biemer, P., Mulry, M., and Parmer, R. (1991), "The Matching Error Study for the 1990 Post Enumeration Survey," Proceeding of the Section on Survey Research Methods, American Statistical Association, 248-253.

78. de la Puente (1993), Why Are People Missed or Erroneously Included by the Census: A Summary of Findings from Ethnographic Coverage Reports. Report Prepared for the Advisory Committee for the Design of the 2000 Census Meeting, March 5. Bureau of the Census, U.S. Department of Commerce.

79. Deming, W. E. (1953), "On a Probability Mechanism to Attain an Economic Balance between the Resultant Error of Response and the Bias of Nonresponse," Journal of the American Statistical Association, **48**, 743-772.

80. Deming, W. E. (1943). Statistical Adjustment of Data. Wiley, New York.

81. Dempster, A. P., Laird, N. M., and Rubin, D. B. (1977), "Maximum Likelihood from Incomplete Data via the EM Algorithm," Journal of the Royal Statistical Society, Series B, **39**: 1-39.

82. Dempster, A. P., and Rubin, D. B. (1983), "The Role of Statistical Theory in Handling Nonresponse," in Incomplete Data in Sample Surveys, Vol. II: Theory and Annotated Bibliographies (W. G. Madow, I. Olkin, and
D. B. Rubin, Eds.). New York: Academic Press.

83. Denham, C., and Rhind, D. (1983), A Census User's Handbook. Edited by David Rhind, Methuen and Co. Ltd, London and New York.

84. Dewdney, J. C. (1983), Census Past and Present: A census user's handbook, pp-1, Edited by David Rhind, Methuen and Co., London and New York.

85. Diamond, I. (1993), "Where and Who are the `Missing Million'? Measuring Census of Population Undercount," Department of Social Statistics, Southampton University, London.

86. Diffendal, G. (1988), "The 1986 Test of Adjustment Related Operations in Central Los Angeles County," Survey Methodology, 14, 71-86.

87. Diffendal, G. J., Schultz, L. K., Huang, E. T. and Isaki, C. T. (1987), "Comparison of Adjustment Methods for Census Undercount in Small Areas," Proceeding of the American Statistical Association, Social Statistical Section, 56-65.

88. Diffendal, Gregg J., Isaki, Cary T. and Malec, Donald J. (1982), "Examples of Some Adjustment Methodologies Applied to the 1980 Census," technical report, U.S. Bureau of the Census, Washington, DC.

89. Drew, J. H. and Fuller, W. A. (1980), "Modelling Nonresponse in Surveys with Callbacks," Proceedings of the Survey Research Methods Section, Journal of the American Statistical Association, 639-642.

90. Drew, J. H. and Fuller, W. A. (1981), "Nonresponse in Complex Multiphase Surveys," Proceedings of the Survey Research Methods Section, Journal of the American Statistical Association, 623-628.

91. Durbin, J. (1954), "Nonresponse and Callbacks in surveys," Bulletin of the International Statistical Institute, **34**: 72-86.

92. Durbin, J. (1958), "Sampling Theory for Estimators Based on Fewer Individuals than the Number Selected," Bulletin of the Statistical Institute **36**: 113-119.

93. Durbin, J., and Stuart, A. (1954), "Callbacks and Clustering in Sample Surveys: An Experimental Study," Journal of the Royal Statistical Society, Series A, **117**: 387-428.

94. Edmonston, B., and Schultze, C. (1995), Modernizing the U.S. Census. Panel on Census Requirements in the Year 2000 and Beyond Committee on National Statistics, National Research Council, National Academic Press, Washington D.C..

95. El-Khorazaty, M. N., Imrey, P. B., Koch, G. G., and Wells, H. B. (1977), "Estimating the Total Number of Events with Data from Multiple Record Systems: a review of methodological strategies," International Statistical Review, 45, 129-157.

96. El-Sayed Nour (1982), "On the Estimation of the Total Number of Vital Events with Data from Dual Collection System," Journal of the Royal Statistical Society, 45, 106-116.

97. Ericksen, E. P. (1974), "A Regression Method for Estimating Population Changes of Local Areas," Journal of the American Statistical Association, 69, 867--875.

98. Ericksen, E. P. (1973), "A Method for Combining Sample Survey Data and Symptomatic Indicators to Obtain Population Estimate for Local Areas," Demography, 10, 137--159.

99. Ericksen, E. P. (1971), "A Method for Combining Sample Survey Data and Symptomatic Indicators to Obtain Population Estimate for Local Areas," Unpublished Ph.D. Thesis, University of Michigan, Ann Arbor, Michigan.

100. Ericksen, E. P., Kadane, J. B., and Tukey, J. W. (1989), "Adjusting the 1980 Census of Population and Housing," Journal of the American Statistical Association, 84, 927-944.

101. Ericksen, E. P. and Kadane, J. B. (1985), "Estimating the Population in a Census year (with comments and rejoinder)," Journal of American the Statistical Association, 80, 98--131.

102. Ericksen, E. P. (1980), "Can Regression Be Used to Estimate Local Undercount Adjustments?" Conference on Census Undercount, Washington, DC: U.S. Government Printing Office, 55-61.

103. Ericson, W. A. (1967), "Optimal Sample Designs with Nonresponse," Journal of the American Statistical Association, **62**, 63-78.

104. Ernst, L. R. (1980), "Variance of the Estimated Mean for Several Imputation Procedures," American Statistical Association, Proceedings of the Survey Research Methods Section, 716-720.

105. Fay, R. E. (1978), "Some Recent Census Bureau Applications of Regression Techniques to Estimation," Presented at the NIDA/NCHS Workshop on Synthetic Estimates, Princeton, New Jersey, April 13-14, Proceeding Published by NIDA, 1979.

106. Fay, R. E., Passel, J. S., and Robinson, J.G., with assistance from C. D. Cowan (1988), The Coverage of Population in the 1980 Census, Evaluation and Research Reports PHC80-E4. Washington, D.C.: U.S. Department of Commerce.

107. Fellegi, I. (1980), "Should the Census Counts be Adjusted for Allocation Purposes: Equity Considerations," Proceeding of the 1980 Conference on Census Undercount, U.S.Bureau of the Census, pp 193-203.

108. Fellegi, I. P., and Holt, D. (1976), "A Systematic Approach to Automatic Edit and Imputation," Journal of the American Statistical Association, **71**, 17-35.

109. Fienberg, S. E. (1992), "Bibliography on Capture-Recapture Modeling With Applications to Census Undercount Adjustments," Survey Methodology, **18**, 143-154.

110. Fienberg, S. E. (1991), "An Adjusted Census in 1990? Commerce Says No," Chance, **4**, 44-52.

111. Fienberg, S. E. (1977), The Analysis of Cross-Classified Categorical Data. Cambridge, Massachusetts: The Massachusetts Institute of Technology Press.

112. Fienberg, S. E. (1972), "The Multiple Recapture Census for Closed Populations and Incomplete 2^K Contingency Tables," Biometrika, **59**, 591-603.

113. Ford, B. L. (1978). Missing Data Procedure: A Comparative Study (part 2). Economics, Statistics and Cooperative Service, U.S. Department of Agriculture.

114. Ford, B. L. (1976), "Missing Data Procedure: A Comparative Study," Proceedings of the Social Statistical Section, American Statistical Association, 324-329.

115. Ford, B. N. (1983), "An Over view of Hot Deck Procedures," In Incomplete Data in Sample Surveys, Vol. II: Theory and Annotated Bibliography (W. G. Madow, I. Olkin, and D. B. Rubin Eds.). New York: Academic Press.

116. Frane, J. W. (1977), "A Note on Checking Tolerance in Matrix Inversion and Regression," Technometrics, **19**, 513-514.

117. Freedman, D.A. (1991), "Policy Forum: Adjusting the 1990 Census," Science, **252**, 1233-36.

118. Freedman, D.A. and Navidi, W.C. (1986), "Regression Model for Adjusting the Census," Statistical Science, **1**, 3-11.

119. Fuller, W. A. (1966), "Estimation Employing Post-strata," Journal of the American Statistical Association, **61**, 1172-1183.

120. Geiger, H., and Werner, A. (1924), "Die Zahl der ion radium ausgesandsen a-Teilhen," Zeitschrift fur Physik, **21**, 187-203.

121. Ghangurde, P. D., and Mulvihill, J. (1978), "Nonresponse And Imputation in Longitudinal Estimation in LFS (Canadian Labour Force Survey)," Household Surveys Development Staff. Statistics Canada Report (Feb. 1978).

122. Ghangurde, P. D. and Singh, M. P. (1977), "Synthetic Estimation in Periodic Household Surveys," Journal of Survey Methodology, Statistics Canada **3**, 152--181.

123. Ghangurde, P. D. and Singh, M. P. (1978), "Evaluation of Efficiency of Synthetic Estimates," 1978 Proceeding of the Social Statistics Section, American Statistical Association.

124. Goldberg, D., Rao, V. R. and Namboodiri, N. R. (1964), "A Test of the Accuracy of the Ratio Correlation Population Estimates," Land Economics, **40**, 100--102.

125. Gonzalez, M. E. (1973), "Use and Evaluation of Synthetic Estimates," Proceeding of the American Statistical Association, Social Statistical Section, 33-36.

126. Gonzalez, E. P. and Hoga, C. (1978), "Small-Area Estimation with Application to Unemployment and Housing Estimates," Journal of the American Statistical Association, **73**, 7--15.

127. Gonzalez, J. F. Jr., and McMillen, M. (1986), "Nonresponse and Noncoverage Analysis in the Southwest Component of the Hispanic Health and Nutrition Examination Survey," American Statistical Association, Proceedings of the Survey Research Methods Section, 326-331.

128. Gonzalez, M. E. and Waksberg, J. (1973), "Estimation of the Error of Synthetic Estimates," Paper Presented at the First Meeting of the International Association of Survey Statisticians, Vienna, Austria, 18-25 August.

129. Goodman, L. A. (1968), "The Analysis of Cross-Classified Data: Independence, Quasi-Independence and Interaction in Contingency Tables With or Without Entries," Journal of the American Statistical Association, **63**, 1091-1131.

130. Gosselin, J. F., and Brackston, G. J. (1978), "The Measurement of Population Undercoverage in the 1976 Canadian Census Using the Reverse Record Check Method," Proceeding of Social Statistics Section, American Statistical Association, 230-235.

131. Greenfield, C. C. (1976), "A Revised Procedure for Dual Systems in Estimating Vital Events," Journal of the Royal Statistical Society, A, **139**, 389-401.

132. Greenfield, C. C. (1975), "On the Estimation of Missing Cell in a 2x2 Contingency Tables," Journal of the Royal Statistical Society, A, **138**, 51-61.

133. Greenless, J. S., Reece, W. S., and Zieschang, K. D. (1982), "Imputation of Missing Values when the Probability of Response Depends on the Variable being Imputed," Journal of the American Statistical Association, **77**, 251-261.

134. Green, M. A., and Stollmack, S. (1981), "Estimating the Number of Criminals," In Models in Quantitative Criminology, (Ed. J. A. Fox), New York: Academic Press, 1-24.

135. Griffin, D. H. and Moriarity, C. L. (1992), "Characteristics of Census Errors," paper presented at the Annual Meeting of the American Statistical Association, Boston, Massachusetts, August 9-13, 1992.

136. Haberman, S. J. (1974), The Analysis of Frequency Data. Chicago: University of Chicago Press.

137. Halacy, D. (1980), 190 Years of Counting America. Elsevier/Nelson Books, New York.

138. Hansen, M. H., Hurwitz, W. N. and Madow, W. G. (1953), Sample Survey Methods and Theory I, John Wiley and Sons, Inc., New York.

139. Hartley, H. O., and Hocking, R. R. (1971), "The Analysis of Incomplete Data," Biometrics, **27**, 783-823.

140. Heady, P., Smith, S., and Avery, V. (1994), 1991 Census Validation Survey: coverage report, OPCS, Social Survey Division, London: HMSO.

141. Herzog, T. N., and Rubin, D.B. (1983), "Using Multiple Imputations to Handle Nonresponse in Sample Surveys," in Incomplete Data in Sample Surveys, Vol. II: Theory and Annotated Bibliography (W. G. Madow, I. Olkin, and D. B. Rubin, Eds.). New York: Academic Press.

142. Hill, C. J., and Puderer, H. A. (1981), "Data Adjustment Procedures in the 1981 Canadian Census of Population and Housing," Current Topics in Survey Sampling (D. Kreweski, R. Platek and J.N.K. Rao, EDs.), 437-454. New York, Academic Press.

143. Hogan, H. (1993), "The Post-Enumeration Survey: Operation and Results," JASA, Vol. 88 **42** pp. 1047.

144. Hogan, H. (1992), "The 1990 Post-Enumeration Survey: An Overview," The American Statistician, **45**, 261-269.

145. Hogan, H. (1992a), "The Post-Enumeration Survey: Operation and New Estimates," paper presented at the Annual Meeting of the American Statistical Association, Boston, Massachusetts, August 9-13, 1992.

146. Hogan, H. (1992b), "New Estimates from the 1990 Post Enumeration Survey," Proceeding of the Survey Research Section, American Statistical Association,

147. Hogan, H. (1989), "Nine Years of Coverage Evaluation Research: What Have We Learned?" Proceeding of the Survey Research Section, American Statistical Association, 663-668.

148. Hogan, H. and Mulry, M. (1987), "Operational Standards for Determining the Accuracy of Census Results," Proceeding of the American Statistical Association, Social Statistical Section, 46-55.

149. Hogan, H. and Wolter, K. (1988), "Measuring Accuracy in a Post-enumeration Survey," Survey Methodology, **14**, 99-116.

150. Holt, D., and Elliot, D. (1991), "Methods of Weighting for Unit Non-response," The Statistician, **40**, 333-42.

151. Holt, D., and Smith, T. M. F. (1979), "Post Stratification," Journal of the Royal Statistical Society, series A, **142(1)**, 33-46.

152. Holt, D., Smith T. M. F. and Tomberlin, T. J. (1979), "A Model-Based Approach to Estimation for Small Subgroups of a Population," Journal of the American Statistical Association, **74**, 405--410.

153. Hughes, P. J. and Choi, C. Y. (1982), Regression Techniques for LGA Population Estimation, Demography and Social Branch Australian Bureau of Statistics, Canberra, 1--24.

154. Hughes, A. L., and Peitzmeier, F. K. (1989), "Weighting and Imputation Methods for Nonresponse in CPS Gross Flows Estimation," American Statistical Association, Proceedings of the Survey Research Methods Section, 279-285.

155. Huggins, R. M. (1989), "On the Statistical Analysis of Capture Experiment," Biometrika, **76**, 133-140.

156. Huggins, R. M. (1991), "Some Practical Aspects of a Conditional Likelihood Approach to Capture Experiments," Biometrics, **47**, 725-732.

157. Isaki, Cary T. (1986), "Bias of the Dual System Estimator and Some alternatives," Communications in Statistics- Theory and Methods, **15(5)**, 1435-1450.

158. Isaki, C. T., Diffendal, G. J. and Schultz, L. K. (1986), "Statistical Synthetic Estimates of Undercount for Small Areas," Paper Presented at the Census Bureau's Second Annual Research Conference.

159. Isaki, C. T., Schultz, L. K., Diffendal, G. J., and Huang, E. T. (1988), "On Estimating Census Undercount in Small Areas," Journal of Official Statistics.

160. Isaki, C. T., Schultz, L. K., Smith, P. J. and Diffendal, G. J. (1985), "Small Area Estimation Research for Census Undercount," in Small Area Statistics: An International Symposium, 219-238, John Wiley and Sons, New York.

161. Isaki, C. T. and Schultz, L. K. (1987), "Report on Demographic Analysis Synthetic Estimation for Small Areas," Statistical Research Division Report Series, CENSUS/SRD/RR-87/03.

162. Isaki, C. T. and Schultz, L. K. (1987a), "Report on the Effects of the Violations of Assumptions on Regression Estimation of Census Coverage Error," Statistical Research Division Report Series, CENSUS/SRD/RR-87/04.

163. Isaki, C. T. and Schultz, L. K. (1986), "Effects of Correlation and Matching Error in Dual System Estimation," Communication in Statistics - Theory and Methods **16(8)**, 22.

164. Isaki, C. T., Diffendal, G. J. and Schultz, L. K. (1987), "Report on Statistical Synthetic Estimation for Small Areas," Statistical Research Division Report Series, CENSUS/SRD/RR-87/02.

165. Jabine, T. B. and Bershad, M. A. (1968), "Some Comments on the Chandra-Sekar-Deming Technique for the Measurement of Population Change," CENTO Symposium on Demographic Statistics, Karachi, Pakistan, November 5th to 12th, 1968.

166. Jackson, C. H. N. (1937), "Some New Methods in the Study of Golossina Morsitans," Procedures of the Zoological Society of London, 1936, 811-896.

167. Jackson, J. E., and Rao, P. S. R. S. (1983), "Estimation Procedures in the Presence of Nonresponse," Proceeding of the Survey Research Methods Section, American Statistical Association, 273-276.

168. Jaro, M. (1989), "Advances in Record-Linkage as Applied to Matching in the 1985 Census of Tampa, Florida," Journal of the American Statistical Association, 414-420.

169. Kadane, J. B., Meyer, M. M., and Tukey, J. W. (1992), "Correlation Bias in the Presence of Stratum Heterogeneity," Technical Report 549, Carnegie Mellon University, Department of Statistics.

170. Kaiser, J. (1989), "The Robustness of Hot-Deck and Cell Mean Methods in Retaining Population Covariance Structure in Imputed Samples," American Statistical Association, Proceedings of the Survey Research Methods Section, 286-289.

171. Kalsbeek, W.D. (1973), "A Method for Obtaining Local Postcensal Estimates for Several Types of Variables," Unpublished Ph.D. Thesis, University of Michigan, Ann Arbor, Michigan.

172. Kalton, G. (1983), Compensating for Missing Survey Data. Research Report Series/Institute for Social Research, University of Michigan, U.S.A.

173. Kalton, G., and Kish, L. (1981), "Two Efficient Random Imputation Procedures," American Statistical Association, Proceedings of the Survey Research Methods Section, 146-151.

174. Kanuk, L., and Berenson, C. (1975), "Mail Survey and Response Rates: A Literature Review," Journal of the Marketing Research 12:440-453.

175. Kish, L. (1990), "Weighting: Why, When, and How?" American Statistical Association, Proceedings Of the Survey Research Methods Section, 121-129.

176. Kish, L. (1980), "Diverse Adjustments for Missing Data," Proceedings of the 1980 Conference on Census Undercount, 83-87, U.S. Bureau of Census, Washington, D.C.

177. Kish, L. (1965), Survey Sampling, John Wiley and Sons, New York.

178. Lauriat, P. (1967), "Field Experience in Estimating Population Growth," Demography, **4**, 228-243.

179. Lee, S. M., and Chao, A. (1991), "Estimating Population Size via Sample Coverage," in proceedings of the 1991 Taipei, Taiwan: Institute of Statistical Science, Academia Sinica, pp. 407-426.

180. Leggieri, C. A. (1994), "Development of a Master Address File as a Base for the 2000 Census," Paper Presented at the Annual Research Conference of the U.S. Bureau of the Census, Arlington, Virginia, March 20-23.

181. Lepkowski, J., and Kalton, G. (1989), "Weighting Adjustments for Partial Nonresponse in the 1984 Sips Panel," American Statistical Association, Proceeding of the Survey Research Section, 296-301.

182. Lincoln, F. C. (1930), "Calculating Waterfowl Abundance on the Basis of Banding Returns," Circular of the U.S. Department of Agriculture, **118**, 1-4.

183. Linder, F. E. (1971), "The Concept and Program of the Laboratories for Population Studies," International Program of Laboratories for Population Statistics, Scientific Series No. 1, Chapel Hill, North Carolina.

184. Linder, F. E. (1970), "A Proposed New Vital event Numeration Unitary System for Developed Countries," Milbank Memorial Fund Quarterly, **48**, 77-87.

185. Lindstrom, H., et al. (1979), "Standard Methods for Nonresponse Treatment in Statistical Estimation," National Central Bureau of Statistics, Sweden.

186. Little, R. J. A. (1989), "Survey Inference with Weights for Differential Sample Selection or Nonresponse," American Statistical Association, Proceedings of the Survey Research Methods Section, 62-72.

187. Little, R. J. A. (1986), "Survey Nonresponse Adjustments for Estimate of Means," Int. Statist. Rev. **54**, 139-154.

188. Little, R. J. A. (1982), "Models for Nonresponse in Sample Surveys," Journal of the American Statistical Association, **77**, 237-250.

189. Little, R. J. A., and Rubin, D. B. (1987), Statistical Analysis With Missing Data. John Wiley and Sons., New York.

190. Lievesley, D. (1983), "Reducing Unit Nonresponse in Interview Surveys," American Statistical Association, Proceedings of the Survey Research Methods Section, 295-299.

191. Luther, N.Y. and Retherford, R.D. (1985), "Consistent Correction of Census and Vital Registration Data," Paper Presented at the Annual Meeting of the Population Association of America, Boston.

192. Madow, W. G., Olkin, I., Nisselson, H., and Rubin, D. B. (1983), Incomplete Data in Sample Surveys, three volumes. New York: Academic Press.

193. Mandell, M. and Tayman, J. (1982), "Measuring Temporal Stability in the Regression Models of the Population Estimation," Demography, 19(1), 135--146.

194. Mantel, N. (1970), "Incomplete Contingency Tables," Biometrics, **7**, 240-246.

195. Marks, E. S., Seltzer, W., and Krotki. (1974), Population Growth Estimation. The Population Council, New York.

196. Marks, E. S. and Waksberg, J. (1966), "Evaluation of Coverage In the 1960 Census of Population Through Case-by-case Checking," Proceeding
of the Social Statistics Section, American Statistical Association,62-70.

197. Martin, J.H. and Serow, W.J. (1978), "Estimating Demographic Characteristics Using the Ratio-correlation Method," Demography, **15**, 223--233.

198. Michaud, S. (1986), "Weighting VS Imputation: A Simulation Study," American Statistical Association, Proceedings of the Survey Research Methods Section, 316-320.

199. Mosher, W., Judkins, D., and Goksel, H. (1989), "Response Rates and Nonresponse Adjustment in a National Survey," American Statistical Association, Proceedings of the Survey Research Section, 273-278.

200. Mulry, M. H., (1991), "Total Error in PEP Estimates for Evaluation Post Strata," 1990 Post Enumeration Survey Evaluation Project P16, unpublished paper.

201. Mulry, M. H., and Dajani, A. (1989), "The Forward Trace Study," Proceeding of the Survey Research Methods Section, American Statistical Association, 675-680.

202. Mulry, M. H., Dajani, A., and Biemer, P. (1989), "The Matching Error Study for the 1988 Dress Rehearsal," in Proceeding of the Section on Survey Research Methods, American Statistical Association, 704-709.

203. Mulry, M. H., and Spenser, B.D. (1993), "Accuracy of the 1990 Census and Undercount Estimates," Journal of the American Statistical Association, 1080-1118.

204. Mulry, M. H., and Spencer, B.D. (1992), "Accuracy of the 1990 Census Undercount Estimates for the Postcensal Estimates," Proceeding of the Survey Research Section, American Statistical Association, 1080-1091.

205. Mulry, M. H., and Spencer, B.D. (1991), "Total Error in PES Estimates of Population," Journal of the American Statistical Association, **86**, 839-863.

206. Mulry, M. H., and Spencer, B.D. (1988), "Total Error in the Dual System Estimator: The 1986 Census of central Los Angeles County," Survey Methodology, **14**, 241-263.

207. Mulry, M. H., and West, K. (1990), "Evaluation Follow-up for the 1988 Post-Enumeration Survey," unpublished paper prepared for presentation at the Annual Meetings of the American Statistical Association, Anaheim, CA, August 5-9.

208. Namboodiri, N.K. (1972), "On the Ratio-correlation and Related Methods of Subnational Population Estimation," Demography, **9**, 443--453.

209. National Research Council (NRC) (1980), Estimating Population and Small Areas, National Academy Press, Washington, DC, pp. 1--247.

210. Nelson, F. D. (1977), ``Censored Regression Models with Unobserved Stochastic Censoring Threshold," Journal of Econometrics **6**:309-327.

211. Nicholls, A. (1977), ``A Regression Approach to Small Area Estimation," Australian Bureau of Statistics, Canberra, Australia, March (Mimeographed).

212. O'Connell, M. A., Bloomfield, P., and Pollock, K. H. (1992), ``Combining the Post Enumeration Survey and Demographic Analysis, a Contingency Table Framework for Adjusting the Census Estimates of Population Size," unpublished manuscript, North Carolina State University, Raleigh, NC.

213. O'Connell, M. A. (1991), "Contingency Table Models for Estimation of the Size of a Partitioned Population," Ph. D. dissertation, North Carolina State University, Raleigh, NC.

214. Oh, H. L., and Scheuren, F. (1978a), "Multivariate Raking Ratio Estimation in the 1973 Exact Match Study," Proceedings of the Survey Research Methods Section, American Statistical Association, 716-722.

215. Oh, H. L., and Scheuren, F. (1978b), "Some Unresolved Issues in Raking Estimation," Proceedings of the Survey Research Methods Section, American Statistical Association, 723-728.

216. Oh, H. L., and Scheuren, F. S. (1983), "Weighting Adjustments for Unit Nonresponse," in Incomplete Data in Sample Surveys, Vol. II: Theory and Annotated Bibliographies (W. G. Madow, I. Olkin, and D. B. Rubin, Eds.). New York: Academic Press.

217. O'Hare, W. (1976), "Report on a Multiple Regression Method for Making Population Estimates," Demography, 13, 369--379.

218 O'Hare, W. (1980), "A Note on the Use of Regression Methods in Population Estimates," Demography, 7, 87--92.

219. O'Muircheartaigh, C. A. and Payne, C. D. (eds) (1977), The Analysis of Survey Data, Volume-2, John Wiley and co., London.

220. OPCS (1996), Census News No. 37, December.

221. OPCS (1996), Census News No. 36, 16 September.

222. OPCS (1993), "Rebasing the Annual Population Estimates," Population Trends No. 73 Autumn 1993.

223. OPCS (1993), "How Complete was the 1991 Census?" Population Trends No. 71 Spring 1993.

224. OPCS (1992), Provisional Mid-1991 Population Estimates for England and Wales and Constituent Local and Health Authorities Based on 1991 Census Results. OPCS Monitor PP1, 92/1, 16 October.

225. OPCS (1992), "Coverage Check From the 1991 Census Validation Survey," Census Newsletter No. 24, 23 October, London.

226. OPCS (1991), Review of Migration Data Sources. OPCS Occasional Paper 39.

227. OPCS (1983), Census 1971, General Report, Part-3 (Statistical Assessment).

228. OPCS (1980), "A Comparison of the Registrar General's Annual Population Estimates for England and Wales, with the Results of the 1981 Census," Occasional Paper {\bf 29}. Population Statistics Division, OPCS.

229. Peterson, C. G. J. (1896), "The Yearly Immigration of Young Plaice into the Limfjord from the German Sea," Report of the Danish Biological Station to the Ministry of Fisheries **6**, 1-48.

230. Platek, R. (1977), "Some Factors Affecting Nonresponse," Bulletin of the International Statistical Institute, **47(3)**, 347-366. Also, Survey Methodology, **3**, 191-214.

231. Platek, R. and Gray, G. B. (1979), "Methodology and Application of Adjustments for Nonresponse," Bulletin of the International Statistical Institute, **48(2)**, 533-570.

232. Platek, R. and Gray, G. B. (1978), "Nonresponse and Imputation," Survey Methodology, **4**, 144-177.

233. Platek, R., Singh, M. P. and Tremblay, V. (1978), "Adjustment for Nonresponse in Surveys," Survey Sampling and Measurement (Namboodiri, N. K. Eds.), Chapter 11. Academic Press, New York.

234. Platek, R. et al. (1987), Small Area Statistics, John Wiley and Sons, New York.

235. Politz, A. and Simmons, W. (1949), "An Attempt to Get the ``NOT-AT-HOME" into the Sample Without Callbacks," Journal of the American Statistical Association, **44**, 9-33.

236. Pregibon, D. (1976), "Incomplete Survey Data: Estimation and Imputation," Methodology Journal of Household Survey Division, Statistics Canada.

237. Pregibon, D. (1977), "Typical Survey Data: Estimation and Imputation," Methodology Journal of Household Survey Division 2: Statistics Canada.

238. Preston, S.; Coale, A.J.; Trussell, T.J.; and Weinstein, M. (1980), "Estimating the Completeness of Reporting of Adult Deaths in Population that are Approximately Stable," Population Studies, **34(2)**, 170-202.

239. Preston, S. and Hill, K. (1980), "Estimating the Completeness of Death Registration," Population Studies, **34**, 349-366.

240 Purcell, N.P. (1979), Efficient Small Domain Estimation: A Categorical Data Analysis Approach. Unpublished Ph.D. Thesis, University of Michigan, Ann Arbor, Michigan.

241. Purcell, N.P. and Kish, L. (1979), "Estimation for Small Domain," Biometrics, **35**, 365--384.

242 Purcell, N.J. and Linacre, S. (1976), "Techniques for the Estimation of Small Area Characteristics," Paper Presented at the 3rd Australian Statistical Conference, Melbourne, Australia, 18-20 August.

243. Pursell, D.E. (1970), "Improving Population Estimates with the Use of Dummy Variable," Demography, **7**, 87--91.

244. Quenouille, M. H. (1949), "Approximate Tests of Correlation in Time Series," Journal of the Royal Statistical Society, B **11**, 68-84.

245. Rao, J. N. K. (1973), "On Double Sampling for Stratification and Analytical Surveys," Biometrika **60**:125-133.

246. Rao, P. S. R. S. and Jackson, J. E. (1984), "Estimation Through the Procedure of Subsampling the Nonrespondents," American Statistical Association, Proceedings of the Survey Research Methods Section, 550-553.

247. Rives, N.W. (1976), "A Modified Housing Unit Method for Small Area Population Estimation," 1976 Proceeding of the Social Statistics Section, American Statistical Association, 717--720.

248. Rockwell, R. C. (1975), "An Investigation of Imputation and Differential Quality of Data in the 1970 Census," Journal of the American Statistical Association, **70**, 39-42.

249. Robinson, J. G., Ahmed, B., Das Gupta, P., and Woodrow, K. (1993), "Estimation of Population Coverage in the 1990 United States Census Based on Demographic Analysis," Journal of the American Statistical Association, 1061-1079.

250. Rosenberg, H. (1968), "Improving Current Population Estimates Through Stratification," Land Economics, **44**, 331-338.

251. Rossmo, D. K., and Routledge, R. (1990), "Estimating the Size of Criminal Population," Journal of Quantitative Criminology, **6**, 293-314.

252. Royce, D. (1992), "Incorporating Estimates of Census Coverage Error into the Canadian Population Estimates Program," Proceedings of the 1992 Annual Research Conference, Bureau of the Census, 18-26.

253. Rubin, D. B. (1987). Multiple Imputation for Nonresponse in surveys. New York: Wiley.

254. Rubin, D. B. (1978), "Multiple Imputations in Sample Surveys," American Statistical Association, Proceedings of the Survey Research Methods Section, 20-34.

255. Rubin, D. B. (1977), "Assignment to Treatment Group on the Basis of a Covariate," Journal of Educational Statistics, **2**, 1-26.

256. Rubin, D.B. (1976), "Inference and Missing Data," Biometrika, **63**, 581-592.

257. Rubin, D. B., Schafer, J. L. and Schenker, N. (1988), "Imputation Strategies for missing Values in Post Enumeration Surveys," Survey Methodology, **14**, 209-221.

258. Sanathanan, L. P. (1972a), "Estimating the Size of a Multinomial Population," Annals of Mathematical Statistics, **43**, 142-152.

259. Sanathanan, L. P. (1972b), "Models and Estimation Methods in Visual Scanning Experiments," Technometrics, **14**, 813-829.

260. Sanathanan, L. P. (1973), "A Comparison of Some Models in Visual Scanning Experiments," Technometrics, **15**, 67-78.

261. Sande, I. G. (1983), Hot Deck Imputation Procedures, in Incomplete Data in Sample Surveys, Vol. III: Symposium on Incomplete Data, Proceedings (W. G. Madow, and I. Olkin, Eds.). New York: Academic Press.

262. Schaible, W.L. (1978), "Choosing Weight for Composite Estimates for Small Area Statistics," Proceedings of the Social Statistics Section, American Statistical Association.

263. Schaible, W.L., Brock, D.B. and Schnack, G.A. (1977), "An Empirical Comparison of the Simple Inflation, Synthetic and Composite Estimators for Small Area Statistics," Proceeding of the Social Statistics Section, American Statistical Association, 1017-1021.

264. Schenker, N. (1988), "Handling Missing Data in Coverage Estimation With Application to the 1986 Test of Adjustment-Related Operations," Survey Methodology, **14**, 87-98.

265. Schenker, N. (1987), "Report on Missing Data in the 1986 Test of Adjustment Related Operations," SRD Research Report Number: CENSUS /SRD /RR-87/09, Statistical Research Division Report Series, U.S. Bureau of the Census, Washington, DC.

266. Schirm, A. L. and Preston, S. H.(1987), "Census Undercount Adjustment and the Quality of Geographic Population Distribution," Journal of the American Statistical Association, **82**, 965--978.

267. Schmitt, C. R.(1952), "Short-cut Methods of Estimating County Population," Journal of the American Statistical Association,**47**, 232-38.

268. Schnable, Z. E. (1938), "The Estimation of the Total Fish Population of a Lake," American Mathematical Monthly, **45**, 348-352.

269. Schultz, L. k., Huang, E. T., Diffendal, G. J. and Isaki, C. T. (1986), "Some Effect of Statistical Synthetic Estimation on Census Undercount of Small Areas," in Proceeding of the Survey Research Methods Section, American Statistical Association.

270. Schultz, L. K., Isaki, C. T., and Diffendal, G. J. (1987), "Report on Using Regression Models for Small Area Adjustment," Statistical Research Division Report Series, CENSUS/ SRD/RR-87/01.

271. Seltzer, W. (1969), "Some Result from Asian Population Growth Studies," Population Studies, **23,** 395-406.

272. Seltzer, W. and Adlakha, A. (1969), "On the Effect of Errors in the Application of the Chandra Sekar-Deming Technique," Paper Prepared for the Population Council Seminar on the Chandra Sekar-Deming Technique: Theory and Application, New York.

273. Shapiro, S. (1950), "Estimating Birth Registration Completeness," Journal of the American Statistical Association, **45**, 261-264.

274. Shapiro, S. (1954), "Recent Testing of Birth Registration Completeness in the United States," Population Studies, 8, 3-21.

275. Shryock, H.S. and Siegel, J.S. (1975), The Methods and Materials of demography, U.S. Government Printing Office, Washington DC.

276. Siegel, J. S. and Zelnik, M. M. (1966), "An evaluation of coverage in the 1960 census of population by techniques of demographic analysis and by composite methods," Proceedings of the Social Statistic Section, American Statistical Association, 71-85.

277. Simmons, W. R. (1954), "A Plan to Account for ``Not-At-Homes" by Combining Weighting and Callbacks," Journal of Marketing 19:42-53.

278. Spar, M. and Martin, J. (1979), "Refinements to Regression-Based Estimates of Postcensal Population Characteristics," Review of Public Data Use, 7(5/6).

279. Spencer, B. (1986), "Conceptual Issues for Measuring Improvement in Population Estimates," Proceedings of the Second Annual Research Conference, U.S.Bureau of the Census, pp 393-407.

280. Spencer, Bruce (1980), "Implications of Equity and Accuracy for Undercount Adjustment: A Decision-theoretic Approach," 204-216 in Proceeding of the 1980 Conference on Census Undercount, Bureau of the Census, Washington, DC.: Department of Commerce.

281. Spencer, B.D. (1985), "Statistical Aspects of Equitable Apportionment," Journal of the American Statistical Association, 80, 815-822.

282. Srinath, K. P. (1971), "Multiphase Sampling in Nonresponse Problems," Journal of the American Statistical Association 66:583-586.

283. Srinivasan, S. K., and Muthiah, S. A. (1968), "Problems of Matching Births Identified from Two Independent Sources," Journal of Family Welfare, 14, 13-22.

284. Starsinic, D.E. and Zitter, M.(1968), "Accuracy of the Housing Unit Method in Preparing Population Estimates for Cities," Demography, 5(1), 475--484.

285. Statistics Canada, (1984a), Postcensal Annual Estimates of Population for Census Divisions and Census Metropolitan Areas, June 1, 1982 (Component Method), Ministry of Supply and Services, Government of Canada, Ottawa.

286. Steffey, D. L., and Bradburn, N. M. (1994), Counting People in the Information Age. Panel to Evaluate Alternative Census Methods Committee on National Statistics, National Research Council, National Academic Press, Washington D.C..

287. Stell, D., and Poulton, J. (1988), "Geographic Estimates of Under-Enumeration," Proceeding of the Section on Survey Research Methods, American Statistical Association, 119-128.

288. Swanson, D.A. and Tedrow, L.M. (1984), "Improving the Measurement of Temporal Change in Regression Models Used for County Population Estimates," Demography, **21(1)**, 373--381.

289. Thomsen, I. and Sirling, E. (1979), "On the Causes and Effects of Nonresponse Norwegian Experiences," In Incomplete Data in Sample Surveys, Vol. III: Symposium on Incomplete Data, Proceedings (W. G. Madow, and I. Olkin, Eds.). New York: Academic Press.

290. Trewin, D. (1977), "The Use of Post-Stratification for Adjustment of Nonresponse Biases," Bulletin of the International Statistical Institute, **47(4)**, 708-712.

291. Tupek, A. R., and Richardson, W. J. (1978), "Use of Ratio Estimates to Compensate for Nonresponse Bias in Certain Economic Surveys," American Statistical Association, Proceedings of the Survey Research Methods Section, 197-202.

292. Tukey, J. W. (1984), "Points to be Made," Presented to the Committee on National Statistics, Panel on Decennial Census Methodology, May 11, 1984.

293. Tukey, J. W. (1981), Discussion of "Issues in Adjusting the 1980 Census Undercount," by Barbara Bailar and Nathan Keyfitz, Paper Presented at the Annual Meeting of the American Statistical Association, Detroit, MI.

294. U.S.Bureau of Census (1988), The Coverage of Population in the 1980 Census, PHC80-E4 Washington, DC.: U.S. Government Printing Office.

295. U.S. Bureau of Census (1985), Statistical Training Document, ISP-TR-5, Washington, D.C.

296. U.S. Bureau of Census (1978), "The Current Population Survey, Design and Methodology," Technical Report No. 40, U.S. Government Printing Office, Washington, DC., pg. 82.

297. U.S. Bureau of the Census (1969), Estimates of the Population Counties and Metropolitan Areas, July 1, 1966: A Summary Report. Current Population Reports, Series P-25, No. 427. U.S. Government Printing Office, Washington, D.C.

298. U.S. Bureau of the Census (1964), "Record Check Studies of Population Coverage," Series ER 60, No.2. U.S. Department of Commerce.

299. U.S. Bureau of the Census (1960), The Post Enumeration Survey: 1950. Bureau of the Census Technical Paper No. 4, Washington D. C.

300. United Nations (1967), "Principles and Recommendations for the 1970 Population Censuses," Statistical Paper Series M44, New York, UN Statistical Office.

301 Verma, R.B.P., Basavarajappa, K.G. and Bender, R.(1984), "Estimation of Local Area Population: An International Comparison, 1984," Proceeding of the Social Statistics Section, American Statistical Association, Washington, DC, 324--329.

302. Warren, R. and Passel, J.S. (1987), "A count of the Uncountable: Estimate of Uncounted Aliens Counted in the 1980 United States Census," Demography, **24**, 375.

303. Wells, H. B. (1971), "Dual Record System for Measurement of Fertility Change," working paper no. **13**, East-West Population Institute, Honolulu, Hawaii.

304. West, S. A., Butani, S. and Witt, M. (1990), "Alternative Imputation Methods for Wage Data," American Statistical Association, Proceedings of the Survey Research Methods Section, 254-259.

305 Whelpton, P. (1950), "Birth and Birth Rates in the Entire United States, 1909 to 1948," Vital Statistics Special Reports, 33, 137 - 162.

306. Whitridge, P. and Kovar, J. (1990), "Applications of the Generalized Edit and Imputation System at Statistics Canada," American Statistical Association, Proceeding of the Survey Research Methods Section, 105-110

307. Witts, J. T., Colton, T., and Sidel, V. W. (1974), "Capture-recapture Methods for Assessing the Completeness of Case Ascertainment when Using Multiple Information Sources," Journal of Chronic Diseases, **27**, 25-36.

308. Witts, J. T., and Sidel, V. W. (1968), "A Generalization of the Simple Capture-recapture Model with Applications to Epidemiological Research," Journal of Chronic Diseases, **21**, 287-301.

309. Wolfgang, G. (1989), "Using Administrative Lists to Supplement Coverage in Hard-to-Count Areas of the Post-Enumeration Survey for the 1988 census of St. Louis," Proceedings of the Survey Research Methods Section, American Statistical Association, 669-674.

310. Wolter, K. M. (1991), "Policy Forum: Accounting for America's Undercounted and Miscounted," Science, **253**, 12-15.

311. Wolter, K. M. (1990), "Capture-Recapture Estimation in the Presence of a Known Sex Ratio," Biometrics, **46**, 157-162.

312. Wolter, K. M. (1986), "Some Coverage Error Models for Census Data," Journal of the American Statistical Association, **81**, 338-346.

313. Wolter, K. M. (1986a), "Capture-Recapture Estimation in the Presence of a Known Sex-Ratio," SRD Research Report Number: CENSUS/SRD/RR-86/20, Statistical Research Division Report Series, U.S.Bureau of the Census, Washington, DC.

314. Wolter, K. M. (1986b), "A Combined Coverage Error Model for Individuals and Housing Units," SRD Research Report Number Census/ SRD/RR-86/27, Statistical Research Division Report Series, Washington, DC: U.S. Bureau of Census.

315. Wolter, K. M. (1984). Introduction to Variance Estimation. New York: Springer-Verlag.

316. Woltman, H., Alberti, N., and Moriarity, C. (1988), "Sample Design for the 1990 Census Post Enumeration Survey," Proceedings of the Section on Survey Research Methods, American Statistical Association, 529-534.

317. Woodrow, K. A. (1992), "A Consideration of the Effect of Immigration Reform on the Number of Undocumented Residents in the United States," Population Research and Policy Review, 11, 117-144.

318. Woolson, R. F. and Cole, J. W. L. (1974), "Comparing Means of Correlated Varieties with Missing Data," Communications in Statistics, **3(10)**, 941-948.

319. Yates, F. (1960), Sampling Methods for Censuses and Surveys, 3rd edition. London: Griffin.

320. Yaukey, D. (1985), Demography: The Study of Human Population, New York.

321. Zaslavsky, A. M. (1993), "Combining Census, Dual-System, and Evaluation Study Data to Estimate Population Shares," Journal of the American Statistical Association, **88**, 1092-1105.

322. Zaslavsky, A. M. (1992), "Combining Census and Dual-system Estimates of population," Proceedings of the 1992 Annual Research Conference, Bureau of the Census.

323. Zaslavsky, A. M., and Wolfgang, G. S. (1993), "Triple-System Modeling of Census, Post Enumeration Survey, and Administrative List Data," Journal of Business and Economic Statistics, **11**, 270-288.

324. Zaslavsky, A. M., and Wolfgang, G. S. (1990), "Triple-System Modeling of Census, Post Enumeration Survey, and Administrative List Data," in Proceeding of the Survey Research Section, American Statistical Association, pp. 668-673.

325. Zelnik, M. (1964), "Errors in the 1960 Census Enumeration of Native Whites," Journal of the American Statistical Association, **59**, 437-459.

326. Zelnik, M. (1965), "An Evaluation of New Estimates of the Negro Population," Demography, **2**, 630-639.

327. Zitter, M and Canavagh, F. J. (1980), "Postcensal Estimates of Population," An Unpublished Paper Presented at the Annual Meeting of the American Association for the Advancement of Science, Session on the 1980 Census, San Francisco, CA, January 1980.

328. Zitter, M. and Shryock, H. S. (1964), "Accuracy of Methods of Preparing Postcensal Population Estimates for States and Local Areas," Demography, **1(1)**, 227--241.

329. Zitter, M., Shryock, H.S., Strasinic, D. and Word, D. (1968), Accuracy of Methods of Preparing Population Estimates for counties. A Summary Compilation of Recent Studies. Unpublished Paper Presented at the Annual Meeting of the Population Association of America, Boston.

Printed in the United States
By Bookmasters